T0174016

Financial Social Innovations

This book helps make sense of the emerging and established social innovations that have disrupted and are disrupting the world of finance. Written in an engaging style, this book offers a systematic study of social innovation in the financial services. It introduces the fundamental concepts of financial social innovations (FINSIs), places them in the context of the broader literature, and provides a new framework for understanding and organising these innovations. The book applies the framework to seven existing FINSIs to illustrate their important components and explore their benefits as well as negative or harmful aspects to society. These seven FINSIs are microfinance, peer-to-peer (P2P) lending, crowdfunding, mobile banking, impact investing, digital cryptocurrencies, and social impact bonds. The easy-to-follow framework will help to ground the reader's understanding of FINSIs as the existing ones evolve and new ones are developed.

This book is ideal for courses on social innovation, social entrepreneurship, and financial innovation in departments of business, economics, social sciences, and political science.

Alessandro Lanteri is Full Professor of Strategy at ESCP Business School, where he teaches executive education programmes. As an expert educator, he helps organisations seize the opportunities of turbulent environments. www.alelanteri.com

Financial Social Innovations

A New Framework to Understand the Social Innovations Disrupting the World of Finance, from Crowdfunding to Bitcoin

Alessandro Lanteri

Routledge
Taylor & Francis Group

LONDON AND NEW YORK

First published 2024
by Routledge
4 Park Square, Milton Park, Abingdon, Oxon OX14 4RN

and by Routledge
605 Third Avenue, New York, NY 10158

Routledge is an imprint of the Taylor & Francis Group, an informa business

British Library Cataloguing-in-Publication Data
A catalogue record for this book is available from the British Library

ISBN: 978-0-367-27657-7 (hbk)
ISBN: 978-1-032-76105-3 (pbk)
ISBN: 978-0-429-29715-1 (ebk)

DOI: 10.4324/9780429297151

Typeset in Times New Roman
by Newgen Publishing UK

This book is dedicated to the many students who have taught me incomparably more than I have taught them.

Contents

About the Author

Alessandro Lanteri is Full Professor of Strategy and Innovation at ESCP Business School, the oldest business school in the world, where he teaches in some of the top-ranked programmes worldwide. He is also the Managing Director of the CLEVER Institute. Furthermore, he teaches in some of the best universities and business schools worldwide, including Hult International Business School in Dubai and New York, and in executive programmes at Saïd Business School (University of Oxford), Cornell University, and London Business School. He has previously held various academic posts at Bocconi University, American University of Beirut, UPO, Grenoble Graduate School of Business, Helsinki University, and UCLA.

An expert educator and advisor Alessandro helps executives, managers, and entrepreneurs navigate turbulent competitive landscapes and respond strategically to the opportunities of the digital age. He is in demand for his expertise in innovation (e.g., design thinking, innovation strategy, business model, social innovation), in digital transformation and the opportunities of emerging technologies (e.g., AI, blockchain, fintech, IoT, the metaverse, robotics, VR), and in strategic foresight (e.g., megatrends, scenario planning, future mindset).

His clients include multinationals, governments, and family businesses across five continents in industries such as capital goods, chemicals, consulting, engineering, financial services, FMCG, logistics, oil & gas, pharmaceutical, retail, technology, telecom, travel and tourism, and utilities. Alessandro also works with international organisations, governments, consulting companies, and start-up programmes. A representative recent selection includes Adani group (India), Asteron (New Zealand), Barilla (Italy), Boston Consulting

Group (Central Europe), Citigroup (USA), CNH Industrial (Italy), Executive Council (UAE), Ford Motors (UK), G20 (Argentina), International Labor Organization (Italy), Ministry of Finance (KSA), Mubadala (UAE), Nissan (France), PWC (Middle East), Sunlife (South East Asia), Unicredit (Central Europe), VISA (Middle East), and World Economic Forum (Switzerland).

Alessandro has authored many international peer-reviewed articles in various journals (e.g., *Journal of Business Ethics, Journal of Social Entrepreneurship, Philosophical Quarterly, Philosophical Studies*) and books published by Cambridge University Press, Palgrave Macmillan, Profile Books, etc. His research has also appeared in *Harvard Business Review* and *MIT Technology Review* issues, *LSE Business Review, World Economic Forum Agenda*, and *Forbes*.

His latest book, *Innovating with Impact*, is published by The Economist/Profile Books (2023). In it, Alessandro and his co-author illustrate the Innovation Pyramid, a strategic process that is rooted in the right cultures and mindsets, leveraging state-of-the-art methods, techniques, and themes to reach the pinnacle of improved performance and value creation. His previous book *CLEVER. The Six Strategic Drivers for the Fourth Industrial Revolution* helps strategic decision-makers navigate disruption and seize the emerging strategic superpowers that give companies a competitive edge in the digital age. *CLEVER* became a number 1 Amazon bestseller in Italy and the United Kingdom in 2019.

Alessandro holds a PhD from Erasmus University Rotterdam and a Master's degree from Bocconi University. He also studied at Harvard University, Said Business School at University of Oxford, Sloan School of Management at Massachusetts Institute of Technology, Stanford University, and Stern School of Business at NYU.

Alessandro has lived and worked in 15+ global capitals in Africa, America, Asia, and Europe. He is based in Turin and Abu Dhabi.

Acknowledgements

By the summer term of my first academic year in London in 2014–15, I taught my first course of social innovation in finance. This book contains many of the ideas and reflections accumulated since. These ideas greatly benefited from conversations with the many brilliant students who took my course since it was first launched and with many outstanding individuals who generously contributed to my courses or to my thinking. It is only fair to mention Stephen Alexander (Consensys), Lorenzo Allevi (Oltre Venture), Vic Arulchandran (Nivaura), Andrea Bonaceto (Eterna Capital), Mario Calderini (Politecnico di Milano), Christopher Dadson (SIB Group), Kieran Garvey (Crowdcube), Mat Gazeley (Zopa), Filippo Giordano (Universita' Bocconi), Guglielmo Gori (SocialFare), Kirsty Grant (Seedrs), Sandrine Henton (Educate Global Fund), Gautam Ivatury (LendLedger, Happy Loans), Daniel Izzo (Vox Capital), Jonathan Jenkins (SIB Group), Ruben Koekoek (ABN Amro), Mark Lamb (Coinfloor), Paolo Landoni (Politecnico di Torino), Genevieve Leveille (AgriLedger), Rebecca MacKenzie (Agora Microfinance), Matteo Marinelli (Maha Agriculture Microfinance), Laura Michelini (LUMSA), Mike Mompi (ClearlySo), Laura Montenegro (SIB Group), Alan Newton (Eventopedia), Laura Orestano (SocialFare), Amadeo Pellicce (Coinfloor), Francesco Perrini (Universita' Bocconi), Karl-Heinz Richter (Engaged Investments), Artur Rtiscev (SIB Group), Alex Soskin (Impact Hub Westminster), Luca Torre (Gawa Capital), Emma Van Dijkum (Founding Circle), and Amy Zawislak (SIB Group). It is quite possible that a few more have undeservedly been left out, for which I feel sorry.

In April 2015, I went back to school, as a learner, at the Oxford Impact Investing Programme at Saïd Business School of Oxford

University. Finally, some of the ideas in this book were presented at the 5th Ashridge International Research Conference in Berkhamsted (UK) in July 2016, at the Financing Social Innovation workshop in Turin (Italy) in October 2016, and at the 77th Annual Meeting of the Academy of Management in Atlanta (USA) in August 2017 and benefited from insightful comments. These opportunities helped me advance and refine my thoughts. Finally, Jeff Raderstrong provided excellent copyediting of this book.

Introduction
Financial Adventures in Beirut

On a warm late afternoon in October 2013, I was in Beirut, sitting on the terrace of a lovely cafe on the Corniche basking in the red sunset over the Mediterranean. As I nervously sipped on my tea, I kept repeatedly checking my pocket and the five $100 bills contained within. I was waiting for a person I knew only by the alias "habibit_961" to purchase something that to my knowledge was not entirely legal.

I had first contacted habibit_961 through an online forum. After I sent my message, I received a mysterious phone call from a person who instructed me in broken English to post on the forum a place and three different times for a potential exchange, before suddenly hanging up. I had come to this cafe twice already during my suggested times, but habibit_961 did not show. The third and last opportunity was just a few minutes from now, and while I waited, I felt equally excited and terrified.

During those few minutes before habibit_961 would (supposedly) show up, I reviewed once again the warnings I had found online. I can still remember one, which said:

> Everyone who has bought or sold bitcoin has been scammed at least once. Make sure to complete the exchange in a public place. The parking lot of a shopping mall is ideal because security will be at hand in case things get ugly.

This terrace of a popular cafe seemed like a good enough public place for me. I waited, looking around at the other customers, wondering if habibit_961 was among them. At around 10 minutes past my final

DOI: 10.4324/9780429297151-1

suggested meeting time, I realised that my mysterious broker was not going to show. I never found out why. Maybe because he or she had planned on scamming me and the public location was too public. Or maybe because habibit_961 was not a real person and I was a part of an elaborate joke. Or maybe whoever was behind the username ended up being too busy to bother with the sale of a few bitcoins.

Had habibit_961 come to the meeting with intentions to make good on our deal, I would have traded my $500 cash for three bitcoins. At the peak value, in November of 2021, those bitcoins would have been worth over $206,000. As of this writing, in early 2023, they would be worth around $87,000 – much lower than the peak, but quite a fantastic return on my original $500 investment. Had I bought them in 2013 and held onto them until their peak price, I would have sold them not to a cash trader like habibit_961 but to one of the many online exchanges that had become accessible and (relatively) reliable in the meantime. Despite the failed transaction, that experience of being stood up at the cafe made quite an impression.

This failed cafe meeting was not the only financial adventure I took back in 2013. While walking home late at night, I found some cash sitting on the sidewalk. Not a soul was to be seen. So, I pocketed the money. The next day, I was teaching my class on business ethics at the American University of Beirut where I was a professor, and I asked my students what they thought I should do with the money. After an intense debate, I resolved to lend it through Kiva, a relatively new organisation at the time, that helped facilitate small loans to entrepreneurs around the world. My class and I chose to lend the money to two Lebanese entrepreneurs: Aline from Tyre, who used the cash I found on the ground to buy new make-up tools for her beauty salon, and Marwan from Nabatieh, who used the loan to make some repairs to his house. After a few months, they paid me back in full and the interest on my loan went to Al Majmoua, a Lebanese non-profit microfinance institution that partnered with Kiva to distribute the funds.

My brush with bitcoin and my class discussion leading to me becoming a micro-lender started me thinking about the changing nature of technology and finance, and how the intersection of the two was influencing our socioeconomic systems. I began to see examples everywhere: Around the same time I was in the cafe waiting to become a bitcoin investor, I had a brief encounter with Abdallah Absi, who had been just chosen as one of Lebanon's top 20 entrepreneurs for his crowdfunding platform, Zoomaal, which is democratising capital access, allowing entrepreneurs to ask for loans directly from individuals.

While in Beirut, I also was involved in many exciting projects in the emerging social innovation ecosystems in Lebanon and Tunisia, thanks to my friends at Beyond Reform and Development, a mission-driven consulting firm. Those were exciting times for social innovation in the Middle East. New funding models, like social incubation and impact investing, were gaining momentum and I was either directly involved in them (Hmayed et al. 2015, Lanteri 2015) or witnessing them from a privileged position (Idrissi 2015, Sfeir 2015, Wyne 2015). For this, I must thank my colleague Dima Jamali who had engaged me in editing two very ambitious volumes describing the evolving landscape of social entrepreneurship across the Middle East (Jamali & Lanteri 2015). Around that time, I was also growing fond of fostering financial inclusion by deploying mobile banking, a model I had heard about from my friends at Nokia in 2009, when I lived in Helsinki.

Despite how much I loved living and teaching in Beirut, I moved to London shortly after my failed bitcoin purchase to start another chapter in my personal and professional life. This move gave me some physical and intellectual space from my recent financial adventures, which allowed me to reflect on them from a new perspective. I began to wonder whether all the anomalous financial models I had witnessed could perhaps be more than just random experiments. Did they have anything in common? Was something major going on in finance which was surfacing in the form of these apparently disconnected new services? In the meantime, London kept offering new fascinating case studies of rapidly evolving financial services, with fintech booming and cryptocurrencies going through the roof.

I began to explore these questions more fully in my new role at the Hult International Business School. I started to teach a course on social innovation in finance, which is where I first started to build out the framework for what this book would become. As I taught the course and continued my research, I could barely keep up with the developments at the intersection of financial innovation and social innovation. Progress was happening too quickly.

If my initial question was "is something major going on in finance?", the answer was clearly "yes". I began to create a framework to help organise and put a structure around these innovations to allow my students to better understand new developments as they were happening. By doing this, we could compare and contrast innovations to allow for critical analysis and deeper exploration of both the process of innovation and the outcome of that innovation.

In the resulting years, the progress continued at a rapid clip. Bitcoin expanded rapidly and, despite a "crypto winter" in 2022 which saw a pricing correction of many crypto assets, the financial innovation that required surreptitious cash exchanges 10 years ago is now an established industry. Microfinance has continued to expand and has made inroads even in developed nations such as the United States and the United Kingdom. Crowdfunding and P2P lending has become more sophisticated and offered access to capital for entrepreneurs who would not be able to grow their businesses without it.

Like many trends, the COVID-19 pandemic accelerated these innovations. Their digital nature helped drive their expansion, and when COVID-19 forced us to rely on digital tools even more, we further sought out these innovations. Bitcoin, for example, surged during the pandemic, and the later pullback in value could be seen as a right-sizing correction acknowledging a COVID-influenced inflation (Jabotinsky & Sarel 2022). Crowdfunding, similarly, increased during the pandemic as people looked for support with groceries, rent, and other necessities (Lerman 2021). While we may not maintain this level of reliance or interest in these innovations, it is clear that my "financial adventures" are no longer one-off escapades, but increasingly a fundamental part of our modern financial and social economic system.

This book helps make sense of emerging and established financial and social innovations that have disrupted and are disrupting our world. It offers a framework to organise these innovations and places them in context of the broader literature. The book then leverages this framework to explore seven existing financial social innovations (or what I will refer to as FINSIs) to illustrate their benefits as well as the negative or harmful aspects on society.

1 Social Innovation, Financial Innovation, and Financial Social Innovation

Social Innovation is a relatively broad concept that incorporates new ideas across activities and fields. Open source in new product development, distance learning in education, integrated area development in urban planning, international labour standards in public policy, low-cost care in healthcare, and fair trade in commerce have all been described as "social innovations" despite the disparate connections between the concepts. Initially celebrated for successfully responding to local social issues, social innovations[1] are increasingly regarded as offering solutions that can be replicated across geographies and scaled to tackle structural and systemic issues (Nicholls et al. 2015a). Social Innovation is now seen as an important solution to the major, global challenges we face as a species: poverty, hunger, climate change, inequality, etc.

Financial Innovation has a narrower field of scholarship, as the concept relates specifically to the field of finance and developments related to the transfer or investment of currencies. Financial Innovation and Social Innovation have points of intersection, and there have been several very successful social innovations within the financial industry. However, the tradition of scholarship on this intersection is relatively nascent, even more nascent than the scholarship of Social Innovation.

This book presents an overview of this scholarship connecting Social Innovation and Financial Innovation and a framework to understand the resulting innovations that operate at the intersection of these concepts. It does so by first introducing the state of the art on Social Innovation research, following the broad review and systematisation conducted by Alex Nicholls at the Saïd Business School

DOI: 10.4324/9780429297151-2

University of Oxford and by his co-authors from both academia and practice (Nicholls et al. 2015b, Nicholls & Murdock 2012b). Next, it introduces the literature on Financial Innovation, following the classification by Josh Lerner at Harvard Business School and Peter Tufano at Saïd Business School. Building on these two streams of research, it proposes a new framework to classify financial social innovations (henceforth FINSIs). The book then explores the new framework through the case studies of seven FINSIs. The last chapter summarises the conclusions and discusses the potential to use the FINSI framework to assess future FINSI.

Incumbent financial companies find it hard to appreciate the potential impact of FINSI because it is emerging from outside the field of finance and because legacy actors look at these advancements with a piecemeal approach, where various FINSIs are treated as unique, scattered phenomena. However, FINSIs are a unified and important phenomenon that have impacted the world and will continue to do so in a variety of ways. The framework presented here will help financial companies, policymakers, academics, and students appreciate the full scope of FINSIs and how we should continue to think about new developments in the area.

The FINSI Framework

The development of social innovations has been universally led by practitioners (Phills et al. 2008). As a consequence, the concept of Social Innovation has been employed very liberally by diverse stakeholders, each emphasising some of Social Innovation's many features depending on their own circumstances, because they valued tangible results and impact rather than conceptual clarity. This fluidity is one of Social Innovation's strengths when it comes to practical applications, because it encourages unconstrained experimentation. However, it can be a limitation for rigorous scholarly investigation.

For scholars, the interdisciplinary and transdisciplinary nature of Social Innovation often means that its "epistemological and methodological stances are in continuous development" (Moulaert et al. 2014, p. 13). Social Innovation is thus best regarded as a "pre-concept" (Chambon et al. 1982) or a "quasi-concept" (Tepsie 2014), adaptable to different contexts and audiences (McNeill 2006). Being adaptable to a variety of domains entails that Social Innovation displays an extraordinarily broad range of applications and conceptualisations. In turn, such variety makes it difficult to develop a singular epistemological approach across domains and disciplines. Indeed, it is sometimes suggested that Social Innovation requires bespoke epistemological approaches for each domain or discipline (Moulaert et al. 2014).

These domain- or discipline-specific epistemologies emerged and consolidated around different research traditions (e.g., design thinking, social entrepreneurship, urban studies), each expanding on its own theoretical roots and refining its own tools. While such fragmentation poses a challenge, it also opens the opportunity to pursue greater understanding within an overarching framework (Crossan & Apaydin 2010). Indeed, attempts at providing an overarching framework have recently encountered success through critical reviews of the different strands of Social Innovation scholarship (Moulaert et al. 2005, 2014, Pol & Ville 2009, Tepsie 2014, Weerakoon et al. 2016, Young Foundation 2012) and subsequent systematisations of this research domain (Cajaiba-Santana 2013, Dawson & Daniel 2010, Nicholls et al. 2015b, Nicholls & Murdock 2012b, Phills et al. 2008). This book moves from such successful efforts and embraces their outcome as a starting point.

Social Innovation

In its most common uses, Social Innovation refers to significant and lasting changes in a society or in its structures, regulations, norms, and values, such that the society's performance in satisfying the economic and social needs of its members is enhanced, at the individual, collective, or system level (Hamalainen & Heiskala 2007, Phills et al. 2008, The Young Foundation 2012, Moulaert et al. 2014, Tepsie 2014, Zollo et al. 2015). With the increasing complexity and interconnectedness of the globalised economy, social innovation is increasingly recognised as a driver of value creation for society and the economy. For example, *Social Economy Science: Transforming the Economy and Making Society More Resilient* is an in-depth study of how social innovation not only creates social outcomes but also strengthens our economy and the institutions on which we rely (Krlev et al. 2023).

In ontological terms, social innovations are "ideas" (Mulgan et al. 2007, Mumford 2002, Nicholls et al. 2015b). These ideas can take different forms: products (e.g., zero-energy housing) and services (e.g., car sharing), processes and procedures (e.g., participatory budgeting), framings (e.g., gender equality), organisational forms (e.g., decentralised autonomous organisations), rules and regulations (e.g., personal budgets), policies and laws (e.g., the U.S. Inflation Reduction Act of 2021), and many more. The examples discussed in this book focus on the *services* form, although they almost invariably encroach on other forms. For example, one of the foundations of microfinance is dismissing the received view that the poor are incapable or lazy, so they will not productively use money lent to them, and that they are opportunistic and untrustworthy, and so will not repay a loan unless some asset is taken as collateral. In other words, besides offering a new *service*, microfinance established a new *framing* of the poor as entrepreneurial and reliable individuals who are eager to improve their station in life.

There are many components of a successful social innovation because of its broad nature. There are many common elements, such as the reliance on cross-sector action, an open and collaborative approach, and a connection to grassroots communities.

Figure 1.1 illustrates the expansive nature of Social Innovation. In the inner concentric circle, you can see elements that are more necessary to a social innovation application. They must have an element of sociality, of course, and be effective in doing so. They also will likely help form new relationships or help with the better use of assets and resources.

Using the FINSI of mobile banking as an example, we see that this has the components of social innovation by increasing financial inclusion by offering unbanked individuals a way to transfer funds. Mobile banking also helps to create new relationships and social connections between those who are sending payments and those who are receiving payments. Mobile banking is also focused on grassroots empowerment, increasing the capacity of local people and offering them new ways to engage with the economy.

There is a broad and very intuitive consensus that two defining elements of Social Innovation are sociality and novelty. These elements have been characterised in very different ways in the literature.[2] Despite this variety, the many different existing scholarly traditions converge around two broad strands of Social Innovation (Pol & Ville 2009, Nicholls & Murdock 2012a, Nicholls et al. 2015b): one emphasises innovation in social *processes* and so captures changes in social interactions, interpersonal relations, and governance models; the second focuses on social *outcomes*, in terms of products and services that meet unmet human needs and address market failures, and

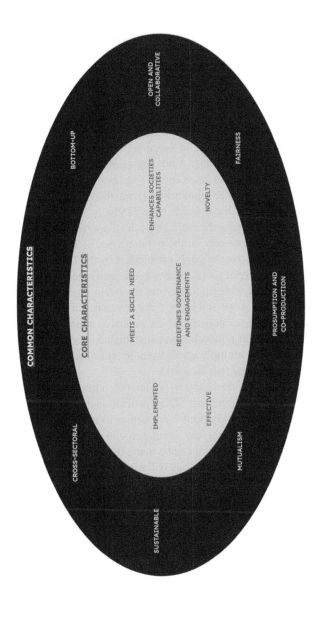

Figure 1.1 Characteristics of Social Innovation.

Source: Adapted from Tepsie (2014).

have a positive impact on the lives of the individuals affected. For example, the bitcoin[3] is a digital currency that is created and exchanged by an open-source network of individuals and organisations, without the involvement or oversight of any central authority. Therefore, it is a social *process* innovation. On the other hand, impact investments are made with the goal of generating both financial returns and positive social impact, like affordable housing and education, so they are social *outcome* innovations.

This category acts as a fundamental threshold: a financial service only counts as a legitimate Social Innovation if it embodies an innovation in sociality as a *process*, an *outcome*, or both.[4] Accepting that social innovations can be either a *process* or an *outcome* is a necessary compromise for Social Innovation being, as mentioned, a "quasi-concept" that requires more scholarship. Using this threshold can help us to better determine what is and is not a FINSI.

This book will leverage the sustainable development goals (SDGs) as an important sub-category of the outcome Social Innovation category. The SDGs are a set of goals established by the United Nations that cut across a variety of social and environmental outcomes – ending poverty, empowering women, improving the environment, etc. These goals are massive in scope and support the creation of a more sustainable economic system, encouraging an inclusive economy and one that is more reliant on renewable and sustainable energy. Achieving the goals will require multiple actors from government, business, and NGOs working together (see Box 1.1).

As discussed in future chapters, social outcomes are difficult to appropriately measure, and the SDGs provide a useful framework for doing so. Many organisations have begun to use the SDGs as a standard way to track progress around social and environmental outcomes. A similar connection can be made between FINSIs and the SDGs, and the remainder of the book will use the SDGs as a grounding mechanism to determine how the FINSI framework interacts with the tangible outcomes produced by each FINSI explored.

In their broad review and systematisation of the existing research, Nicholls and colleagues (Nicholls et al. 2015b, Nicholls & Murdock 2012b) also identify three levels and four dimensions of Social Innovation. The levels refer to the extent of change the innovations bring about. At one extreme, there are *incremental* innovations that improve on existing goods and services, to increase their efficiency or effectiveness. For example, P2P lending allows

Box 1.1 The Sustainable Development Goals

In September 2015, the General Assembly of the United Nations (UN) adopted the resolution "Transforming Our World: The 2030 Agenda for Sustainable Development" (United Nations 2015). This 2030 agenda contains an ambitious vision and a plan of action to make the world a better place. But what does a "better place" mean? It means a place where individuals can increasingly choose what is best for themselves, without being limited by needs or insufficient liberties.

This approach is based on Nobel Laureate Amartya Sen's capabilities approach (Sen 1993, 1999) according to which human development consists in the authentic freedom for individuals to genuinely choose what they value and have a reason to value. The capabilities approach underpins our current understanding of human development and global political ethics.

The 2030 agenda consists of a set of 17 integrated sustainable development goals (SDGs), further divided into 169 individual targets. Ban Ki-moon, secretary-general of the United Nations at the time when the agenda of 2030 was launched stressed the importance of the SDGs by declaring that they "are our shared vision of humanity and a social contract between the world's leaders and the people" (Ki-moon 2015).

These are the SDGs:

Goal 1. End poverty in all its forms everywhere.

Goal 2. End hunger, achieve food security and improved nutrition and promote sustainable agriculture.

Goal 3. Ensure healthy lives and promote well-being for all at all ages.

Goal 4. Ensure inclusive and equitable quality education and promote lifelong learning opportunities for all.

Goal 5. Achieve gender equality and empower all women and girls.

Goal 6. Ensure availability and sustainable management of water and sanitation for all.

Goal 7. Ensure access to affordable, reliable, sustainable, and modern energy for all.

Goal 8. Promote sustained, inclusive, and sustainable economic growth, full and productive employment, and decent work for all.

Goal 9. Build resilient infrastructure, promote inclusive and sustainable industrialisation and foster innovation.

Goal 10. Reduce inequality within and among countries.

Goal 11. Make cities and human settlements inclusive, safe, resilient, and sustainable.

Goal 12. Ensure sustainable consumption and production patterns.

Goal 13. Take urgent action to combat climate change and its impacts.

Goal 14. Conserve and sustainably use the oceans, seas, and marine resources for sustainable development.

Goal 15. Protect, restore, and promote sustainable use of terrestrial ecosystems, sustainably manage forests, combat desertification, and halt and reverse land degradation and halt biodiversity loss.

Goal 16. Promote peaceful and inclusive societies for sustainable development, provide access to justice for all and build effective, accountable, and inclusive institutions at all levels.

Goal 17. Strengthen the means of implementation and revitalise the global partnership for sustainable development.

Achieving the SDGs creates the conditions for individual freedom, because it entails eliminating those conditions that prevent individuals from pursuing what they care about and have a reason to care about. For example, poverty forces people to compromise on their choices, in order to survive; poor health prevents them from following through with their decisions; lack of education impairs their ability to make well-informed decisions; inequality blocks access to valuable choices; and so on.

From the perspective of Social Innovation, the SDGs offer a very important framework to assess social outcomes. A FINSI will be considered to have a social outcome when it promotes the achievement of 1 of the 169 targets of the SDGs. For many of these targets, it will be difficult to make a direct linkage between an innovation and the outcome itself, as the SDGs were conceived of on a country-wide scale. However, they offer a useful starting point when considering what elements of social and financial innovations lead to tangible social outcomes.

efficient and rapid loans to individuals and enterprises, although often they could have obtained the loan from elsewhere (Baek et al. 2014). Instead, *institutional* innovations reconfigure entire social and economic structures to create new forms of value. For example, mobile banking modified the existing market structure, initially in Kenya and later in several other African economies. Finally, *disruptive* innovations pursue radical or large-scale system change that often overthrow established hierarchies and power relations.[5] Such is the case of cryptocurrency, which has created a completely new financial industry. When social innovations are classified according to the dimension of their impact, Nicholls and Murdock (2012b) list four types: *individual*, *organisation*, *network* or *movement*, and *system*. For instance, while P2P lending affects the individuals involved, crowdfunding impacts organisations, Bitcoin engages a network, and microfinance affects entire systems. These elements just reviewed constitute the basic building blocks for a framework to analyse Social Innovation.

Three additional categories are often discussed in the literature and are therefore worth including: who is the innovator, what are the drivers pushing the innovation, and what are the functions of a particular social innovation. Consistently with the variety of its manifestations, Social Innovation happens in all sectors: public (government), third sector (nonprofit or NGO), private and informal (Murray et al. 2010, Phills et al. 2008, Ramani & Vivekananda 2014). Similarly, social innovations perform a range of functions and are driven by a multitude of factors. The advent of globalisation, combined with the accelerated development and spread of technology, has enabled people to address issues that were previously beyond their reach. Moreover, changes in demography due to migration or ageing, unemployment rates, etc., also resulted in social issues that inspired various social innovations. The COVID pandemic forced a great restructuring of our society, leading to new social innovations and the acceleration of the adoption of existing ones. Many other factors have been observed to drive social innovation, such as the limitations of the government-driven welfare and development models, market changes, growing inequality, and more. Although all these forces and changes contribute to explaining the appearance and development of FINSIs, this book more specifically explores the drivers behind financial innovations and their functions.

Financial Innovation

A broadly accepted definition describes Financial Innovation as "the act of creating and then popularising new financial instruments as well as new financial technologies, institutions and markets" (Tufano 2003, p. 310). There is also a consensus that the critical element to understanding financial innovations is their function (Crane et al. 1995, Merton 1992, Tufano 2003). Financial systems need to meet a range of societal demands (Lerner & Tufano 2011). To fulfil such demands, financial products and services perform six basic functions:

- moving funds across time and space (e.g., savings accounts);
- pooling funds (e.g., mutual funds);
- managing risk (e.g., insurance and many derivatives products);
- extracting information to support decision-making (e.g., markets that provide price information, such as extracting default probabilities from bonds or credit default swaps);
- addressing *moral hazard* and *asymmetric information* problems (e.g., contracting by venture capital firms); and
- facilitating the sale or purchase of goods and services through a payment system (e.g., cash, debit cards, credit cards).

The emergence of individual financial innovations can be traced back to six main drivers: completing inherently incomplete markets; addressing inherent *agency problems* and *information asymmetries;* minimising *transaction, search or marketing costs;* responding to taxes and regulation; responding to increasing globalisation and risk; and responding to technological shocks. For example, equity crowdfunding allows small investors to acquire risk capital in nontraded companies, and so make speculative investments that would otherwise be impossible (*incomplete markets*). Microfinance institutions use group lending to manage *asymmetrical information* and reduce *transaction costs*. Digital cryptocurrencies emerged, thanks to innovative *technologies* and in response to *regulations* and increased *risk*.

It is easy to see how these functions and drivers differ from those of other (and perhaps more conventional) social innovations. This raises the question of whether the functions and the drivers of FINSIs correspond to a relevant notion of sociality of *processes* and *outcomes*. As mentioned earlier, the sociality of *processes* is particularly manifest in those innovations that are made possible by vast networks of interacting parties (e.g., Bitcoin) or radically modify the interactions between network users (e.g., P2P lending) by means of *disintermediation*. As for the sociality of *outcomes*, they often involve *financial inclusion* (Allen et al. 2012, Demigurc-Kunt & Klapper 2012) by means of giving access to financial services that were previously unavailable for some parties (e.g., microfinance) or did not exist altogether (e.g., crowdfunding) or by better supporting social purpose organisations (e.g., impact investing). These two dimensions of sociality can be recognised by looking at the functions and drivers of financial innovations (Lerner & Tufano 2011). It is precisely because their drivers and functions are different from those of other social innovations that studying FINSIs as an autonomous subfield within Social Innovation is warranted – and necessary.

The FINSI Framework

The following framework helps to fully crystallise FINSIs as the intersection between Financial and Social Innovation. The framework to analyse FINSIs is built around seven categories identified so far and summarised in Table 1.1. The first category, which tracks innovation form (product, service, framing...), is not fully explored in this book because, as previously mentioned in the overview of Social Innovation, FINSIs are services. The next category is type (process or outcome), which, as discussed, is a critical threshold and the litmus test for what is and is not a social innovation.

The next two categories – level (disruptive, incremental, or institutional) and dimension (individual, organisation, network, or system) – classify the scale and type of social impact generated. Sector (informal, private, public, or third sector) is explicitly included because the most effective and original social innovations are often cross-sectoral (Murray et al. 2010, Phills et al. 2008) or

Table 1.1 The FINSI Framework

Form	Type	Level	Dimension	Sector	Financial functions	Drivers
Products and services	Process	Incremental	Individual	Informal	Moving funds	Incomplete markets
Processes & procedures	Outcome	Institutional	Organisation	Private	Pooling funds	Asymmetric information
Framings		Disruptive	Network system	Third public	Managing risk	Transaction costs
Organisational forms					Extract information	Regulations
Rules and regulations					Asymmetric information	Globalisation/Risk
Policies and laws					Payment	Technology

originate outside their traditional sectors. The last two categories map the critical elements of Financial Innovation: financial function (moving funds, pooling funds, managing risk, extracting information, addressing moral hazard and asymmetric information problems or payment) and driver (incomplete markets, asymmetric information, transaction costs, technology, regulations, or risk).

The two streams of literature reviewed previously map the fundamental characteristics required to understand Social Innovation and Financial Innovation and usefully classify social and financial innovations. The new framework combines these two streams into a broader tool for FINSI. The remainder of this book will demonstrate how this new framework is useful to classify FINSIs.

Seven FINSIs

We can use this framework to classify existing and emerging FINSIs. The FINSIs featured in this book meet four criteria of (1) being a *process* social innovation, an *outcome* social innovation, or both; (2) operating in the financial sector; (3) being in existence for at least 5 years; and (4) being present in some form in more than one country. The first two conditions are self-explanatory in the light of the discussion of the framework. The third condition was imposed to ensure the identification of social innovations that had endured a test of their long-term viability. Finally, the fourth condition was imposed to filter out local phenomena that possibly reflect unique contextual conditions and may be unfit for scaling out to other contexts. If Financial Innovation means "creating and popularising new financial instruments", criteria (3) and (4) jointly serve as a test of popularisation. Both (3) and (4) were verified at the time when this research started in 2015.

The FINSIs included in the present study (Table 1.2) met all of these criteria. The research investigated each model drawing from secondary sources and published academic research.

Each of these FINSIs will be explored in depth in the coming chapters, but in summary they are

- Microfinance: A process and outcome innovation in which small amount of money is loaned to entrepreneurs and individuals using alternative due diligence and collateral requirements to support traditionally unbanked and underinvested opportunities.

Table 1.2 Seven FINSIs

Financial social innovation	Representative examples	Main references
Microfinance	Agora (UK), Bandhan (IN), Blue Orchard (CH), Compartamos (MX), Grameen Bank (BD), PerMicro (IT)	Armendariz and Morduch (2010), Cull et al. (2009)
P2P lending	Funding Circle (UK), Lending Club (US), Prosper (US), Faircent (IN), Smartika (IT), Zopa (UK)	Baek et al. (2012), Chen et al. (2014), Wei & Lin (2016)
Crowdfunding	ArtistShare (US), CrowdCube (UK), Exporo (DE), Ketto (IN), Kickstarter (US), Seedrs (UK)	Agrawal et al. (2014), Belleflamme et al. (2014), Belleflamme and Lambert (2014), Vulkan et al. (2016)
Mobile (branchless) banking	bKash (BD), M-Pesa (KE), Wizzit (ZA)	*The Economist* (2012), Jussila (2015), Rangan and Lee (2012)
Impact investing	Acumen (US), Bridges Ventures (UK), OltreVenture (IT), Social Investment Group (UK)	Daggers and Nicholls (2016), Global Impact Investing Network (2022), Wilson (2015)
Digital cryptocurrencies	Bitcoin (US), Ethereum (CH), Litecoin (US)	Ali et al. (2014a, 2014b), European Central Bank (2012)
Social impact bonds	Massachusetts (US), Peterborough Prison (UK), Rotterdam Municipality (NL), Utkrisht bond (IN)	Gustafsson-Wright (2020), Liebman and Sellman (2013), Nicholls and Tomkinson (2015)

- P2P lending: A process innovation where individuals are able to receive investments or loans directly from other individuals rather than financial institutions.
- Crowdfunding: A process innovation that allows a group of individuals – sometimes hundreds or thousands – to collectively pool resources to support a project or initiative.
- Mobile (branchless) banking: An outcome innovation in which banks and other financial institutions can engage and/or connect with customers to send payments through their mobile phone or other digital tools rather than a traditional physical branch location.
- Impact investing: An outcome innovation in which traditional financial products and tools are applied to achieving social results.
- Digital cryptocurrencies: A process innovation that creates fully digital currencies that can be bought and exchanged through online platforms and networks without third parties.
- Social impact bonds: A process and outcome innovation in which outcome-based contracts promote specific social results during a set time period.

All of these FINSIs meet the first criteria of supporting sociality through a *process* or *outcome*, the key gating criteria for Social Innovation and FINSIs. Some of them encourage connection with others through financial interactions (e.g., P2P lending, crowdfunding) and some create direct outcomes improving the social system (e.g., impact investing, social impact bonds). Some have innovative approaches to both processes and outcomes (e.g., microfinance). Most FINSIs will not have clear dividing lines between the framework categories, but instead have degrees of each dimension.

For example, in the system I used to lend small loans to Lebanese entrepreneurs mentioned in the introduction, Kiva operates as a platform that facilitates P2P, crowdfunded, microfinancial services to entrepreneurs in developing countries disbursed by private citizens in developed countries. This organisation leverages many of the FINSIs explored in this book to operate effectively. The Lebanese nonprofit organisation Al Majmoua that managed my loans through Kiva is a more traditional microfinance institution (MFI), which offers financial services to low-income individuals and entrepreneurs. For this reason, studying microfinance as a FINSI, even with the intersection with other FINSIs in practice, affords greater clarity. Therefore, this research focuses on Social Innovation models rather than the case studies of the individuals or the organisations that initiated them. Doing so allows a deeper understanding of the mechanisms underlying the phenomenon under study (Phills et al. 2008). It also allows identifying characteristics and drawing conclusions that are more general and not implementation specific.

These FINSIs should also not be considered an exhaustive list. In fact, the opposite. These FINSIs represent an illustrative list of innovations at this present moment which have developed enough to establish a level of scholarship and examples to be appropriately assessed within the framework. The subsequent analysis in this book should be seen as a test case for future work around FINSIs and illustrate how the framework should be used as FINSIs develop in greater numbers.

Practical Implications

An in-depth validation of the theoretical implications of this framework is included in Appendix 2. The main focus of this research was the development and validation of the framework proposed. Therefore, the results discussed concern its overall viability and not its specific theoretical connections. Exploring its content in greater detail is a clear direction for future research.

The validation and exploration of the framework in this book's subsequent chapters points to a major practical finding: All of these FINSIs were first developed by nonfinancial institutions. This finding might be due to a selection problem, or a failure to identify other FINSIs that could have been included in this study and that were launched by traditional financial institutions. It is therefore possible that this common feature will be disproved in the future. For the time being, however, and to the extent that FINSIs do originate outside financial institutions, they are also instances of the so-called alternative financial services or AFS (Baek et al. 2014). AFS are "financial channels and instruments that emerge outside of the traditional financial system (i.e., regulated banks and capital markets)" (Cambridge Centre for Alternative Finance 2016). FINSIs are a subclass of AFS.[6] However, there are good reasons to maintain the two categories separated. AFS are defined as a "negative" category of financial services not initially created by mainstream financial institutions. Yet, many mainstream financial institutions are now offering these services. Furthermore, the institutions offering such financial tools are progressively being assumed under the control of financial authorities and so they are effectively becoming financial institutions. It is somewhat unclear whether these services remain "alternative" under these conditions. For example, would cryptocurrencies issued by a central bank (Bech & Garratt 2017, Camera 2017, Monetary Authority of Singapore & Deloitte 2017) count as AFS?

Another practical implication is that certain phenomena are best understood in the aggregate. For example, the 1987 Black Monday, the 2008 financial crisis, the 2015 Chinese stock market crash, and the massive declines in stock prices at the onset of the COVID-19 pandemic occurred when the price of many stocks rapidly dropped at the same time. One approach to understand these crashes would be to analyse the reasons why each investor decided that each of these stocks should be sold at that particular moment in time. Perhaps it would be possible to track all these explanations and successfully aggregate them into a higher-level explanation. However, this effort would most likely be frustrating and inefficient. Market crashes are best understood as aggregate phenomena, with peculiar causes and characteristics, rather than the composite effect of individual stocks dropping in price. Aggregate phenomena display "emergent properties" (O'Connor & Wong 2015), which are distinct and different from the properties of its components. A classic example is that of water, which is composed of two molecules of hydrogen and one molecule of oxygen. The human body is over 60% water, with individual organs like the brain and lungs peaking at over 70% and 80%, respectively. Yet, the properties of hydrogen and oxygen do not explain the properties of the human body or even those of water. Similarly, market crashes are best understood as aggregate.

Interestingly, pundits are precisely taking this aggregate level approach when they discuss the emergence of financial technologies or *fintech*. The growth and development of FINSIs have coincided with a similar explosion in the growth of fintechs. The global fintech market is expected to reach $1.5 trillion by 2030, with growth especially strong in the emerging markets where many FINSIs have been deployed for improved social outcomes (Boston Consulting Group 2023). Fintech is not interesting because individual venture capitalists invest in separate start-ups deploying selective technologies to address some niche in particular geographies. Instead, the consensus is that a broader phenomenon is taking place: Digital technologies are being systematically leveraged and deployed to transform financial services. This calls for a broader outlook of the phenomenon. I invite the readers to consider that the same is in order for FINSIs.

Like with AFS, FINSIs should be considered a separate category from fintech despite the overlap. While many FINSIs leverage fintech digital technologies for the delivery of their financial functions, there is a distinct difference between the two. Fintechs do not inherently create sociality, although they can. Mobile apps have democratised stock trading, for example, without improving social outcomes or increasing social connections. FINSIs, similarly, can leverage technology for their social outcomes but do not have to. Many microfinance institutions leverage technologies like mobile banking to support their operations, but core to the microfinance approach has been the "social collateral" element, developed long before digital technology was commonplace (Postelnicu et al. 2014).

As explored in the coming chapters, traditional financial institutions are taking notice of the emergence of FINSIs. However, their response is generally limited to specific services or subsets of their business operations. This is very likely an inadequate response. FINSIs are not just scattered experiments, but different manifestations of the underlying phenomenon of Social Innovation, which is rapidly changing the landscape of financial services. Incumbent financial institutions should understand better these new financial services and what drives their growth and success and design a comprehensive response strategy. In order to do so, they need a theory. Clayton Christensen (Harvard Business Review 2012) put it clearly: "data is only available about the past", therefore "the only way you can look into the future... [is] to have a good theory". The proposed framework is the first, if crude, theory about financial social innovations.

The Dark Side of FINSIs

While this book embraces the dominant and positive outlook on Social Innovation, it is fair to acknowledge some of its drawbacks, both conceptual and practical (Larsson & Brandsen 2016). While market failures are the natural domain for Social Innovation (Austin et al. 2006, Santos 2012), the dominant logic underpinning most social innovation initiatives is one inspired by market mechanisms (Lanteri & Perrini 2021), and aimed at ensuring welfare initiatives are financially sustainable and independent from dwindling public funding.

Innovations that do not scale are treated as failures. Yet, social innovators do not always need or desire to scale (Larsson & Brandsen 2016). The pursuit of financial sustainability through earned income is regarded as a key achievement to demonstrate success. Yet, social innovators do not always want to run a business that earns income (Parkinson & Howorth 2008) and doing so often results in mission drift.

In a case study of the failure of a social enterprise, the Seedco Policy Center (2007, pp. 1–2) explored how a focus on financial innovations can lead to detrimental outcomes. It

> found that nonprofits driven to meet a double bottom line [...] have far more typically led to frustration and failure, drawing attention and resources away from the organisation's core work – and that even the oft-cited success stories are less cut-and-dried than they appear.

This focus on the double bottom line can cause negative social outcomes despite the promise of positive results. There is a longstanding debate, for example, around microfinance and its ability to effectively lift the unbanked out of poverty. There are hundreds of cases of borrowers granted loans they cannot repay, who end up in a permanent state of debt, which eventually makes them into outcasts or even lead them to commit suicide (Hulme & Maitrot 2014). One of the most celebrated microfinance institutions, Banco Compartamos of Mexico, achieved an astonishing growth by charging interests in excess of 100%, which eventually led to widespread backlash (Carrick-Cagna & Santos 2009).

Acknowledging this "dark side" of FINSIs, the subsequent chapters will not only look at the positive developments but also explore the negative externalities and/or challenges with implementing the highlighted FINSIs.

FINSIs Go Mainstream

The remaining chapters of this book will explore each of the seven FINSIs in more detail, highlighting their growth and scale in the recent years. The markets for these innovations have recently exploded as more and more traditional financial institutions see their value and advances in technology allow easier adoption (Table 1.3).

Table 1.3 Overview of Seven FINSI's

Financial social innovation	Start date	Market size
Microfinance	1970s	$180 billion[a]
P2P lending	2005	$82 billion[b]
Crowdfunding	2006	$1.6 billion[c]
Mobile banking	2007	$55 billion[d]
Impact investing	2007	$1.1 tillion[e]
Digital cryptocurrencies	2009	$1 trillion[f]
Social impact bonds	2010	$500 milion[g]

Notes
[a] Arti et al. (2021).
[b] Acumen Research and Consulting (2023).
[c] Grand View Research (2023), equity numbers only.
[d] Custom Markets Insights (2022), mobile payments only.
[e] GIIN (2022).
[f] Czervionke et al. (2022).
[g] Gustafsson-Wright and Painter (2023).

The chapters will also explore the "dark side" of these innovations, illustrating how FINSIs are tools to be leveraged that can just as easily lead to positive outcomes as well as negative. They will also illustrate the positive aspects of the FINSI, connecting these outcomes to the SDGs, recognising that some FINSIs have more direct ties to the SDGs than others. The chapters will conclude with a discussion of future implications of each FINSI.

Notes

1 Following prevalent use, this book uses capital letter Social Innovation to refer to the overarching, abstract concept, and to small letter social innovation(s) as the manifestations or applications. The same happens with Financial Innovation and financial innovation(s) and with Financial Social Innovation and financial social innovation(s).
2 For a comprehensive review, the reader should consult Nicholls et al. (2015b), Tepsie (2014), and the sources referred to therein.
3 Bitcoin with a capital B is the protocol for the exchange of the cryptocurrency bitcoin with a small b.

4 These two strands describe sociality as process or outcome. Innovation, too, can be described as process or outcome (Crossan & Apaydin 2010). Innovation could also refer to the diffusion of, and value created by, an outcome (Phills et al. 2008). This book follows the state of the art in focusing on the *process/outcome* classification of sociality only.

5 In this sense "disruptive" differs from Clayton Christensen's concept of disruptive innovation (Christensen et al. 2015).

6 Alternative finance includes financial services, like SME mini-bonds and pension-led funding, which are arguably not FINSIs.

2 Microfinance

After having been initially developed in different locations in the 1970s, microfinance rose to fame through the success story of Prof. Muhamed Yunus and his Grameen Bank in Bangladesh, winning the Nobel Peace Prize in 2006. Microfinance consists of the provision of credit and other financial services on a micro-scale, mostly to *unbanked* or unbankable individuals who would otherwise be unable to access it (Armendariz & Morduch 2010, Cull et al. 2009, Milana & Ashta 2012). Most microfinance institutions (MFIs) offer small loans, also known as microcredit, and for a long time, microfinance was frequently conflated with microcredit. However, microcredit involves only loans, whereas microfinance refers to a broader range of financial products, including savings or insurance.

Poor and unemployed people do not have the income, the assets, and often even the personal identification documents required to do business with commercial banks. They are limited in their ability to accumulate any savings and access credit. For example, when Agora Microfinance started its operations in Zambia in 2011, only 37.3% of the 14.5 million population had an account with a financial institution and seven rural regions had no bank at all.

Lacking access to formal financial services means that when the unbanked save, insure themselves, or borrow, they do so through informal and inefficient channels (Armendariz & Morduch 2010). For example, they participate in rotating credit schemes and burial societies and borrow from other unbanked relatives who can only spare small amounts or from loan sharks who charge exorbitant interest rates. These inefficient financial services make it hard for the unbanked to make the investments necessary to ensure a higher income. So,

DOI: 10.4324/9780429297151-3

they are caught in a vicious cycle of poverty. Microcredit helps self-employed and small entrepreneurs escape this vicious cycle, accelerate the growth of their businesses, and promote self-sufficiency (see Box 2.1).

Traditional commercial lenders screen borrowers to ensure they have an income sufficient to repay the loan and require some assets as *collateral*, which they can seize in case of default. Microcredit, however, is targeted at borrowers who cannot demonstrate a regular income and often do not own assets. This entails major *transaction costs*: identifying the right borrowers for which very little information is available (*asymmetric information*), providing them with micro-loans as low as $10, sometimes in inaccessible locations scattered across vast rural regions, and finally enforcing repayments in the absence of collaterals (*moral hazard*). These costs have been overcome through two major innovations contained within microfinance: *group lending* and *dynamic incentives*. These two innovations perform many financial functions, chiefly *extracting information* and managing *asymmetric information*, and make it possible to create a worldwide credit market for the poor (*incomplete market*).

Box 2.1 The Vicious Cycle of Poverty and the Multiplier Effect of Microfinance

Among the main causes of persistent poverty in less advanced economies are the obstacles to the accumulation of capital (financial and human). This is because of a vicious cycle of forces that act and react in a way that keeps countries poor (Nurkse 2009). The same forces operate at the household level. For example, a family with low income will have low savings. Low savings only allow low investment, in physical and productive assets as well as in education. As a consequence, future income will remain low, therefore perpetuating this condition.

One of the main motivations for microfinance is disrupting the vicious cycle of poverty, by triggering a positive multiplier effect (Figure 2.1). The provision of credit for investment in valuable assets will help families earn higher income. A higher income makes it possible to save more money and to invest in further assets. Such assets will generate a yet higher income, therefore reinforcing this virtuous cycle. In this way, the effect of the initial investment on income is multiplied.

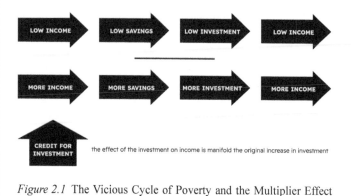

Figure 2.1 The Vicious Cycle of Poverty and the Multiplier Effect
of Microfinance.

Group Lending

Microfinance loans are not offered to individual borrowers but to
groups whose members are jointly responsible for the repayment.
This way, the costs of asymmetric information are passed onto the
borrowers who are better positioned to identify and group with reliable co-borrowers (what is called "assortative matching"), whose loans
they voluntarily take responsibility for. Group lending also reduces the
problem of moral hazard for the lender, because more people can be
held responsible for repayment and because this stimulates a system of
peer monitoring and social sanctioning. The size of these microfinance
groups varies, ranging from 5 borrowers in the Grameen Bank model
(the pioneer of microfinance in Bangladesh) up to 30 in the FINCA
model (an international MFI). Aggregating borrowers into groups
also creates economies of scale that reduce some of the transaction
costs associated with the disbursement and collection of micro-loans
and repayments. Offering and servicing a single loan of $10 might be
unsustainable for a bank, but offering 30 such loans to a group adds up
to $300 or more, a more manageable sum to service and monitor. Often
several group loans are disbursed and managed together, creating further efficiencies. Group lending is therefore a *process* innovation.

Dynamic Incentives

MFIs also create dynamic incentives to repay by not disbursing loans
all at once. Instead, a fraction of the required loan is offered to only

some group members on each of frequent meetings (often weekly) and repayments for these loans begin immediately. This way, groups receive the full loan amount only if they show the ability to pay it back right from the beginning, regardless of the success of the investment for which the loan was given. Moreover, the frequent instalments ensure that borrowers pay back their loans before they have the opportunity or the temptation to spend money otherwise. Similar to group lending, dynamic incentives are also a *process* innovation.

The model of microfinance described here is the traditional version of microcredit. There now exist innumerable variants (e.g., without group lending) and extensions to this model. MFIs are increasingly offering savings accounts and affordable micro-insurance products. Sometimes, these products are bundled with credit. For example, some lenders set aside part of the regular loan repayments in a savings account, which becomes a collateral in case of default. After the borrower repays the load, he or she can access the accumulated savings (Armendariz & Morduch 2010).

The FINSI Framework

Table 2.1 illustrates the different categories of the FINSI framework as they relate to microfinance. They are as follows:

- Main social innovation: Process and Outcome. Microfinance represents both a process and outcome social innovation in that it creates innovations in how finance is delivered, including through the group lending technique and creating dynamic incentives for repayment. It also serves as an outcome innovation because it expands access to banking to an underserved market, empowers women, and helps build financial stability.
- Level: Disruptive. Microfinance is disruptive because it is a large-scale change within a system (finance) and helps to rework power dynamics between the poor and the wealthy.
- Dimension: System. Because of its disruptive qualities, microfinance operates at the systemic level.
- Sector: Private and Third Sector. Microfinance operates within the private markets, as well as in the third sector or nonprofit/NGO space. Most MFIs operate as third sector nonprofits or NGOs, but increasingly traditional financial institutions are offering microfinance products.

Table 2.1 Microfinance

Financial social innovation	Main social innovation	Level	Dimension	Sector	Financial functions	Main drivers
Microfinance	Process Outcome	Disruptive	System	Private (third)	Extract information Asymmetric information	Incomplete markets Transaction costs Asymmetric information

- Financial functions: Microfinance performs the financial functions of extracting information and managing asymmetric information. The group lending technique helps to extract more information about which individual would be a safe or worthy investment, as well as manage asymmetric information because the groups have more information about their peers than a financial institution.
- Main drivers: Microfinance is driven by incomplete markets, transaction costs, and asymmetric information. Many individuals in poverty are unbanked and microfinance supports reaching this market. Transaction costs are quite high for this market due to the small amounts of loans required, and asymmetric information makes it difficult to identify valuable investments, and microfinance helps mitigate these challenges.

Growth and Scale

The total microfinance market is valued at around USD 180 billion and is projected to grow around 10% annually each year to reach almost USD 500 billion in 2030 (Arti et al. 2021). This is a substantial increase since 2009, when the market was about USD 75 billion (Convergences 2019). In urban areas across emerging economies, MFIs are becoming quite common. This growth is also due to the increasing adoption of microfinance by traditional financial institutions, as well as advances in technology that allow the proliferation of microfinance through other FINSIs, such as mobile banking or P2P lending. Traditional banks are projected to be as large of a provider of microfinance products as MFIs by 2030 (Arti et al. 2021).

However, rural areas remain underserved. This is especially problematic in countries where agriculture remains prevalent. For example, 70% of Myanmar's workforce is in agriculture, but most of them only have access to informal financial services, many of which are unreliable or unnecessarily expensive (for more on alternatives to microfinance, see Box 2.2). Maha Agriculture Microfinance offers services that are precisely targeted at farmers, to help them buy agricultural inputs, like fertilisers and seeds, to increase their crop yields and income. In 2022, they served over 64,000 clients, a sixfold increase over the last four years (Maha 2022).

Box 2.2 Informal Financial Services in Emerging Economies

Microfinance represents an alternative to limited money management and investment opportunities for poor individuals living in emerging economies. These individuals have very limited options for managing their assets, which are frequently held in cash and valuables like jewellery, which is ripe for exploitation, theft, and abuse.

These options include

- Day-to-day cash management: Individuals can carry cash, which has the benefit of full liquidity, but has a risk of expropriation and does not allow for borrowing. Rotating savings and credit associations (ROSCAs) are informal lending circles, in which a group of people contribute to and withdraw from a common fund. These have no fees and strengthen a feeling of community but also have a high potential for loss and inflexible terms based on the volatility of other members' needs. Finally, a bank transaction account can securely store funds, but these are difficult to access and have high fees.
- Accumulating lump sums: Individuals can hide cash, but this method has a high risk of loss and will lose value through inflation. ROSCAs, and similar programmes like Accumulating Saving & Credit Associations (ASCAs), are other opportunities to accumulate funds, but carry similar risks of loss. Bank savings accounts can support accumulations of funds, but again are inaccessible for many and have low returns.
- Coping with risks: Poor individuals in emerging economies have little options to reduce their risk from unexpected life events, besides reducing their consumption or selling assets. ROSCAs and moneylenders can help support during times of unexpected costs, but moneylenders in particular can have high fees and threaten safety. Insurance and emergency credit are inaccessible to most because of their high premiums and unpredictable claims processes.
- Transferring money: Delivering money, including remittances, is generally done through in-person interactions, which can be slow and risky, and if you must do so by bus

or taxi or other means of transportation, this can be costly. Mobile payments (discussed in Chapter 5) offer an alternative, as they are low risk, fast, and cheap but do require a mobile phone.

The Dark Side of Microfinance

After years of "good press" for microfinance, the world saw the dark side of this FINSI in the early 2010s. Between 2010 and 2012, stories emerged of individuals and entrepreneurs connected to MFIs who committed suicide, exposing microfinance to more scrutiny. The Government of India, for example, found that in 2010 alone, more than 200 indebted people in the province of Andhra Pradesh killed themselves. The Indian government blamed MFIs for pushing people to suicide, sometimes relying on debt collectors to seek repayments and seize as basic of assets as silverware. The Associated Press, through reporting and document analysis, confirmed that debt collectors for a major MFI in the region intimidated borrowers and in some instances encouraged clients to seek prostitution to repay loans or kill themselves if they could not repay (Associated Press, 2012).

These outcomes go against one of the central promises of microfinance of lifting the poor out of poverty. Unfortunately, these challenges are not unique to India, nor have they stopped after the stories of the Andhra Pradesh suicides came to light. In 2022, Bloomberg completed an expansive report on the state of the microfinance industry and its massive growth in the past decade. It looked at microfinance across the globe and found concerning practices within the industry:

In Cambodia, the average loan provided by so-called microfinance institutions has ballooned sevenfold over the past decade to about $4,200, almost three times the country's average household income, data compiled by the Cambodia Microfinance Association show. Women there have been pressured to sell their homes to repay loans, according to human rights groups and academics who have studied the matter. In Jordan, one of the few countries that still imprisons people for nonpayment of debt, more than 23,000 women were wanted by the police in 2019 for owing less than $1,400 each, Justice Ministry officials have said. In Sri Lanka, consumer-advocacy groups estimate 200 women indebted to microfinance companies committed suicide in the past three years.

(Finch and Kocieniewski 2022)

The dark side of microfinance illustrates a central finding of the FINSI framework: FINSIs are tools of financial and social innovation, but the outcomes they produce are not predetermined. If implemented incorrectly, they can cause disastrous consequences, in the same way predatory loan sharks can hurt families and harm communities. Microfinance should be implemented as a FINSI with integrity and standards to minimise harm. For example, CGAP, the international NGO, has developed a set of principles for microfinance that were endorsed by the G8 developed nations that supports a holistic approach to microfinance (CGAP 2004):

1 The poor need a variety of financial services, not just loans.
2 Microfinance is a powerful instrument against poverty.
3 Microfinance means building financial systems that serve the poor.
4 Financial sustainability is necessary to reach significant numbers of poor people.
5 Microfinance is about building permanent local financial institutions.
6 Microcredit is not always the answer.
7 Interest rate ceilings can damage poor people's access to financial services.
8 The government's role is as an enabler, not as a direct provider of financial services.
9 Donor subsidies should complement, not compete with private sector capital.
10 The lack of institutional and human capacity is the key constraint.
11 The importance of financial and outreach transparency.

Implications for Social Outcomes

The goal of microfinance is lifting its beneficiaries from poverty by means of financial inclusion. Its impact, however, extends beyond the direct effects on borrowers. Successful entrepreneurs create jobs that further benefit their communities. Lending to women, as many MFIs do, helps empower them. These benefits slowly accrue to further stimulate inclusive growth and promote human development.

Because of the disruptive and systemic influence, the resulting social outcomes from microfinance are hard to quantify and define. Using the sustainable development goals (SDGs) introduced in Chapter 1, microfinance could reasonably influence or impact each of the goals or targets. There are obvious connections to Goal 1 (ending poverty), but there could also be a connection to Goal 14 (supporting ocean life) if an entrepreneur is able to use microfinance loans to create more sustainable practices within her fishing business, for example.

Some MFIs explicitly track their impact on the SDGs, so we can at least understand the impact of specific MFIs on the goals and see anecdotal evidence of this FINSI's overall effect on global sustainability. Maha Agriculture Microfinance in Myanmar, discussed earlier, contributes to three goals: Goal 2 (ending hunger), Goal 5 (gender equality), and Goal 8 (decent work and economic growth). Their 2022 impact illustrates the following results towards these goals: Over 64,000 farmers used their services, supporting greater food production and stability. Over half of their staff are women, leading to greater gender equality, and they employ over 250 people in good jobs offering decent work (Maha 2022).

FINCA, the international MFI, and its portfolio contribute to 10 of the 17 SDGs. They and their partners in countries around the world support the SDGs through things like quality education through student loans and savings service for tuition assistance or school supplies. Almost half of their global borrowers are women, supporting female empowerment. In countries where women's rights are limited, FINCA spends more resources ensuring their products reach the women who need them. They have also developed a clean energy loan that directly supports the financing of clean and renewable energy (FINCA 2019).

Finally, the BBVA Microfinance Foundation is a philanthropic (third sector) initiative operating across Latin America with various MFIs. They have had an impact on 13 of the SDGs and have been able to quantify the dollar amounts dedicated to those goals. Over USD 700 million has gone to gender equality, for example, around USD 160 million to sustainable cities, and USD 100 million to peace and justice. The United Nations has commended BBVA for its contributions to the SDGs, and BBVA has supported international partnerships around the goals, including with UN Women to support low-income women entrepreneurs in Latin America and the Caribbean (BBVA Foundation).

Example: Opportunity International[1]

Like Grameen Bank, Opportunity International was an early pioneer of microfinance. Started in the mid-1970s, the organisation was founded by a group of business leaders who wanted to find a solution to "intractable poverty that plagued many nations". These business leaders completed a needs assessment to understand the unique challenges of the global poor, and this study determined that the biggest opportunity was offering a stable income and a path to self-sufficiency.

With this information, Opportunity International then worked to create a model that would support the poor in building their own self-sufficiency. They focused on financial products, given that traditional banks saw these individuals as unqualified borrowers. Instead of focusing only on microcredit, as Opportunity International grew in size and scale, the organisation has developed models for three different areas of microfinance:

- Savings: People living in poverty frequently have no place to store their money or assets, which leaves them vulnerable to thefts and greater insecurity. Opportunity International offered savings options, giving those in poverty a solid foundation to support greater financial security.
- Insurance: Similar to savings, insurance for those in poverty is often overlooked, even by MFIs. Opportunity International developed the "MicroEnsure" product in 2005 to offer protection against flooding, drought, hospitalisation, and other unforeseen tragedies, slowly scaling this model and gaining additional investors.
- Training: Like many MFIs, Opportunity International offers its borrowers more than just loans, but also financial literacy and other financial capacity building programmes. These trainings come through one-on-one relationships with counsellors who meet with borrowers at frequent touchpoints.

Opportunity International almost exclusively works with women, as women are usually the main caretaker of the family and better at managing family finances as compared with0 men. Of their almost 19 million clients, 97% are women. They have loaned out nearly USD 3 billion across 33 countries, which are home to about three quarters of those across the world who live in extreme poverty (Opportunity International 2022).

Future Trajectories

Microfinance has grown from its humble beginnings to create a new system of financing for a previously underserved population. As the microfinance industry has scaled, it has reached a level of stability that will continue to grow and change, interacting with other FINSIs and financial innovations as new ones develop. MFIs and, increasingly, commercial financial institutions, continue to offer additional services to clients, expanding to areas discussed in this chapter, including savings, insurance, and other types of loans. Many other FINSIs, like crowdfunding, P2P lending, and mobile payments are able to leverage the innovations of microfinance to support an expansion of services and offerings.

Central to the evolution and future of microfinance will be the interplay and interaction with fintech and similar digital technologies. As digital engagement becomes the norm, and mobile phones are cheaper and more accessible, fintech opportunities will help to support the expansion of microfinance offerings, gaining new customers and entering new markets. However, fintech does not only support increased accessibility but also supports new innovations to refine the microfinance model. For example, fintech can help capture user data and determine what products customers need, offering more targeted services. Social media footprints can also be leveraged by credit-scoring algorithms to determine the potential reach of a business, helping to appropriately size a loan (Knowledge at Wharton, 2018).

Microfinance was originally created to support those "at the bottom of the pyramid" – essentially the poorest of the poor in developing countries. As microfinance has continued to expand, a "missing middle" has arisen that cannot access traditional financing but for which microfinance loans are too small. Those in this middle are small- to medium-sized enterprises (SMEs) that employ staff and would need something larger than what an MFI could offer but do not have enough collateral to work with a traditional lender (Knowledge at Wharton, 2018). This missing middle is also driving microfinance to more developed countries as an alternative to traditional finance, such as the Community Development Financial Institution (CDFI) sector in the United States. Crowdfunding, discussed in Chapter 4, is also emerging as a microfinance-like approach to supporting SMEs. While much of the products and services are beyond the original conception of microfinance, the sector will continue to grow to meet the needs of those unbanked and to which traditional finance is inaccessible.

Finally, MFIs will continue to expand their scope to focus more on the holistic needs of an entrepreneur or individual. Often, there are needs of an individual that cannot be met by microfinance loans alone but serving those needs will help with repayments and gaining a return on investment. For example, Opportunity International is working to build toilets and help communities access clean water, which creates a stronger foundation for a thriving economy, as well as supporting the creation of good schools to encourage educational attainment (Knowledge at Wharton, 2018). As more and more commercial providers begin to offer microfinance products, the traditional NGO or third-sector MFIs can expand their scope to more holistic solutions to poverty.

Note

1 Based on Bullough et al. (2015).

3   Peer-to-Peer (P2P) Lending

Peer-to-peer (P2P) lending is the practice of individuals making *unsecured loans* to unrelated individuals or organisations without going through a traditional financial intermediary such as a bank or other traditional financial institution (Baek et al. 2014, Chen et al. 2014, Culkin et al. 2016, Wei & Lin 2016). This lending takes place online on peer-to- peer lending companies' websites using various different lending platforms and credit-checking tools. Most P2P loans are unsecured personal loans. (Secured loans are sometimes offered by using luxury assets such as jewellery, watches, vintage cars, or fine art as collateral.) Like all FINSIs, P2P lending interacts with many other FINSI models, such as microfinance and crowdfunding. Microfinance leverages some of the P2P lending innovations, whereas P2P lending can sometimes be considered a type of crowdfunding.

These loans can be for personal purposes, as well as for business purposes. Of course, the basics of P2P lending have been around since the beginnings of our market system. People have always loaned to each other in unsecured terms, without the support of financial institutions. Friends have supported friends in business endeavours, family has supported family in getting through hard times. Advances in *technology* are what helped expand the use of P2P lending, create a market, and establish it as a true FINSI.

Early pioneers of P2P lending like Zopa and Funding Circle began around 2005 to leverage the nascent connectivity of the internet to facilitate a broader expansion of the FINSI. Since then, as technology continues to advance and individuals are more comfortable operating in a digital space, P2P lending has grown in popularity. P2P services are cheaper, faster, and more transparent than incumbents, such as a

DOI: 10.4324/9780429297151-4

credit card company for a borrower or a savings deposit account for the lender. A borrower can get a lower interest rate than a credit card company from P2P lending, and a lender can get a higher return than from a savings account. P2P lending is generally offered only to very reliable borrowers to minimise the risk of default and costly funds recovery, who would have been able to obtain funds elsewhere. This makes fees lower. Other benefits to borrowers include greater flexibility and speed, as well as no requirement for collateral. Additional benefits to lenders include flexibility, user-friendly platforms, and a diversification from the stock market (Baek et al. 2012).

P2P lending is a *process* FINSI of disintermediation, where borrowers and lenders engage through entirely new models of business and interpersonal relations, often in direct communication with each other and with other borrowers and lenders in online communities. This process innovation is driven by *technology* and involves *pooling funds* from lenders and *moving funds* to borrowers, without the inter-mediation of financial companies. These functions are made possible through two innovations: *web-based platforms* that connect peers and in the form of *algorithms* for the analysis of big data used to determine interest rates.

Web-Based Platforms

Digital platforms allow for seamless connections between peers regardless of geography or locations. These platforms open up a P2P network beyond an individual's actual "peers", allowing him or her to access a group of people at a scale never before possible in the history of finance. This innovation mirrors the expansive nature of other digital tools and technologies, with the most obvious corollary being social media. The platforms themselves are a selling point for many borrowers and lenders, with many citing user-experience-related criteria like "ease of use", "customer experience", "speed", "transparency" as a driver for using P2P lending (Baek et al. 2014).

Interest Rate Algorithms

Beyond the relatively pervasive innovation of web-based platforms, P2P lending also leverages algorithms to facilitate their lending products and reduce transaction costs of lending. These algorithms rely on machine learning to make investment decisions, and increasingly

P2P lending firms are looking to artificial intelligence (AI) to support their lending decision-making (Ariza-Garzón et al. 2021).

For example, Zopa and Funding Circle both developed proprietary credit scoring systems, which assign risk profiles based on many more variables than traditional banks. In particular, Zopa found that the owners of a certain brand of mobile phones are slightly less likely to default on a loan than owners of another brand and can therefore be charged marginally lower interest rates.

There are two different typical algorithmic approaches to setting an interest rate for P2P lending: posted price and reverse auction.

Posted Price

- A borrower will post a loan with a duration, amount, and purpose.
- The algorithm will automatically check the credit score of the individual, setting an interest rate based on the score and other factors incorporated into the algorithm. Borrowers with higher credit scores are charged lower rates.
- Lenders will then choose to fund the loan in full, or multiple lenders can pool their money together to fund the loan.

Reverse Auction

- A borrower will post a loan with a duration, amount, purpose, and the maximum interest rate they will accept.
- Lenders will privately list the minimum rate they will accept for a loan.
- Multiple lenders will put out a bid on the loan.
- The algorithm then determines who will fund the loan by choosing the lender(s) bidding the lowest interest rate. The lenders then pay out based on either the interest rate they accepted OR the maximum interest rate offered by the borrower (Chen et al. 2016, Wei and Lin 2016).

The choice of pricing mechanism is crucial because it affects the operations of the market and its overall outcomes. For the posted price mechanism of determining interest rates, loans are more likely to be funded, the time to complete the funding is shorter, and the average sum lent is higher. However, for the reverse auction mechanism, interest rates will be lower and there will be less probability of default (Wei & Lin 2016).

The FINSI Framework

Table 3.1 illustrates the different categories of the FINSI framework as they relate to P2P lending. They are as follows:

- Main social innovation: Process. P2P lending is an innovation in disintermediation, removing the need for traditional financial institutions. It allows for direct connection between borrowers and lenders, increasing communication and engagement between previously unconnected actors.
- Level: Incremental. P2P lending offers small efficiencies as compared with traditional financial institutions in the form of lower fees and increased transparency.
- Dimension: Individual and Organisation. P2P lending operates at the level of the individual or organisation, supporting personal loans as well as business loans, mostly for small- and medium-sized enterprises (SMEs) (Katsamakas & Sánchez-Cartas 2022).
- Sector: Informal and Private. P2P operates within the private markets as lending platforms, but due to the disintermediation, can also operate within the informal economy for those without access to traditional financing options.
- Financial functions: P2P lending performs the financial functions of moving funds and pooling funds. Funds are moved between individuals and organisations, and pooled when multiple lenders choose to support a single loan.
- Main drivers: P2P lending, in its modern context, is driven by technology. While P2P lending has occurred across history to provide financial services to those without access to traditional banking, the FINSI P2P lending iteration has been made possible by increasing digital access and digital connectivity (Berdnorz 2023).

Growth and Scale

The P2P lending market is around USD 82 billion and expected to grow to over USD 800 billion by 2030, growing at close to 30% each year. Consumer lending accounts for about 80% of this market overall, which includes personal loans, student loans, and mortgages. The high growth is due to accelerating use of technology as well as increasing

Table 3.1 P2P Lending

Financial social innovation	Main social innovation	Level	Dimension	Sector	Financial functions	Main drivers
P2P lending	Process	Incremental	Individual Organisation	Informal Private	Moving funds Pooling funds	Technology

demand for alternative financing options (Acumen Research and Consulting 2023).

This growth and market size comes as a contrast to the state of P2P lending when it first began. At their outset, P2P lending platforms were seen as relatively risky or even "shady". During the 2008 financial crisis, P2P lending grew in popularity due to the broad dissatisfaction with traditional financial institutions, and the general limiting of credit that forced individuals and small businesses to look for capital elsewhere (Berdnorz 2023). A similar acceleration occurred during the COVID-19 pandemic in 2020 as many interactions shifted to a digital-first space (Najaf et al. 2022). However, because the pandemic led to loss of employment and general decreases in income across the globe, many P2P borrowers were unable to repay their loans and some lending platforms had to pause lending activities (Acumen Research and Consulting 2023).

The Asian markets once represented a major portion of the P2P lending industry, with the P2P market share peaking around 2017 with the majority of P2P loans happening in China. Because of the low levels of penetration with financial institutions, many borrowers looked to P2P lending to meet their financial needs. However, some borrowers misrepresented themselves and were unable to pay back their loans. (See "The Dark Side of P2P Lending".) The Chinese P2P lending market declined in 2020 and has not recovered (Deng 2022, Suryono et al. 2019). North American and European markets currently make up most of the market share of P2P lending (Acumen Research and Consulting 2023).

The Dark Side of P2P Lending

The volatility described in the previous section highlights the challenges with P2P lending. Borrowers have an incentive to over-represent their ability to repay the loan, which creates a "moral hazard" in which borrowers have more information about the potential success of a loan as compared with the lender. Deng (2022) finds P2P lending as similar to that of the "Lemons" market of Akerlof (1970), exploring the market for used cars:

> As such, the P2P lending market is analogous to the "Lemons" market described in Akerlof (1970), in which the equilibrium is that borrowers of poor quality drive out those of high quality (Gresham's law). Therefore, the only reason why borrowers with

good repayment capability turn to P2P lending is that they do not have free access to formal finance.

(Deng 2022)

This leads to a "crowding out" within the market. As more formal financing options grow, high-quality borrowers will seek out those opportunities rather than P2P lending. This leaves greater levels of low-quality borrowers within the P2P market (Deng 2022).

This has led to the algorithmic lending described previously, which prioritises high-quality borrowers with good credit scores and proven reliability. However, relying more and more on algorithms and machine learning can also lead to discrimination. In particular, machine learning with P2P lending has led to gender discrimination, with women more likely to be funded than men (Suryono et al. 2019). Machine learning has also been shown to lead to racial discrimination in certain cases (Turner Lee 2018).

Finally, P2P lending remains a generally unregulated market, with challenges of appropriate regulation in a rapidly changing industry driven by technological advances (Deng 2022, Suryono et al. 2019). There is an ongoing debate about how to best regulate P2P lending and solve for some of the moral hazard and discrimination concerns, either through self-regulation or a broader regulatory framework (Basha et al. 2021). This debate mirrors a similar concern about regulating the cryptocurrency industry, discussed further in Chapter 7.

Implications for Social Outcomes

Because the focus of P2P lending is an innovation within a process, there is a greater impact on intangible connections and the movement of funds rather than delivering a specific outcome. However, P2P lending has been implemented in ways that can produce an outcome, linked to the sustainable development goals (SDGs). Arguably, P2P lending could support a broad range of the SDGs for the same reasons that microfinance does. P2P lending offers financial products and services to those who traditionally have not had access to financial institutions. This can lead to expanded financial inclusion within a population or network (Katsamakas & Manuel Sánchez-Cartas 2022), which has all the potential benefits discussed in the microfinance chapter: lower poverty rates, increased equality between genders, higher standard of livings, educational access through student loans, etc.

However, the potential for P2P lending to create positive social outcomes relies on the level of interest rate offered for loans. The highest social impact from P2P lending occurs when the loans given are interest free. These interest-free loans offer the greatest opportunity for social outcomes and can support women's empowerment, for example (Dorfleitner et al. 2021).

Example: Lending Club[1]

Lending Club was an early pioneer of P2P lending, and the first to register with the US Securities and Exchange Commission. It was founded by Renaud Laplanche in 2007, who recognised that the difference between the incredibly high credit card interest rates of 18% and low rates on bank deposits offered an opportunity for an entrepreneur. He was a software engineer and thought there could be a way to leverage digital technologies to solve this problem.

Lending Club grew rapidly in its first years. Between its founding and 2016, it facilitated over USD 13 billion in loan originations. The majority of Lending Club's users used its loans to consolidate credit card debt. Of the USD 930 billion in credit card debt in the United States, around half met Lending Club's credit models. This meant that users could significantly reduce the interest paid on debt through P2P lending. Lending Club offered two main products: standard loans and custom loans. The standard loans were three-to-five-year unsecured loans. The custom loans included small business loans, student loans, and other personal loans. Borrowers would frequently use the Lending Club loans to consolidate and pay off credit card debt, which could have rates as high as 17%. Lending Club offered lenders a relatively high interest rate of 8% and an opportunity to diversify their portfolio.

Lending Club decided to go public in 2015. It offered initial shares at USD 15 for its initial public offering (IPO) but was quickly oversubscribed. Its initial market valuation was almost USD 6 billion. In Q4 of 2015, Lending Club was facilitating nearly USD 2.6 billion in loan originations. Over the lifetime of the company, that number was around USD 13.4 billion.

After Lending Club's initial IPO, their stock price began to fall. Within a few months, the stock price was at about 50% of the IPO price. Competitors began to enter the market, encouraged somewhat by Lending Club's example, but Lending Club remained the only public company and had greater restrictions to be able to adapt to new competitors. As more and more companies began to compete for Lending Club's customers, it became more and more difficult for the company to operate and perform as it had done previously.

On top of these additional competitors and scrutiny, Lending Club's challenges continued, with a financial mismanagement scandal that led to the resignation of Laplanche in 2016. Laplanche had failed to disclose a personal stake in a fund that Lending Club was considering an investment in, and also mishandled funds of a sale of near-prime loans to a single investor. The founder and CEO was replaced by the then-president. The executive team determined that there was nothing inherently wrong with the company or its business and proceeded as planned without Laplanche at the helm.

Despite the executive team's assertions, Lending Club lost USD 145 million in 2016. They brought on Valeria Kay that same year to head their institutional investing group and help turn around the company. She focused her work on packaging consumer loans for institutional investors, which helped support the evolution of the company. Lending Club eventually pivoted away completely from P2P lending and decided to shift focus to institutional investing (Balogh 2019).

Future Trajectories

All of the recent volatility in the P2P lending space makes it difficult to project how the industry will evolve. Like Lending Club, Zopa, another early pioneer of P2P lending in the United Kingdom, has completely shut down its P2P services, focusing on growing its banking and credit card business. The company cited lack of profitability and declining interest from retail investors in P2P lending during the COVID-19 pandemic as major factors for influencing the shift (Lanyon 2021). Zopa was actually the first major P2P lending firm, so the company's shift may be a sign of things to come.

The decline of China's P2P lending market, the pivots of Zopa and Lending Club, and P2P's general "crowding out" challenges may point to P2P lending gradually fading out, despite projections of future growth (Acumen Research and Consulting 2023). Like Lending Club's growth and ultimate stumbles, P2P lending may be a victim of its own success. Pioneers of P2P lending were able to prove that a market for some kind of alternative financing existed, and that there was interest in retail investing from consumers. Those that were operating within a "shadow banking" informal environment shifted towards P2P lending, and with the success of this FINSI, are now shifting into more traditional financial banking opportunities.

Other FINSIs and fintech have begun to offer additional products to consumers, meeting the financing needs that P2P lending once provided. Fintechs alone make financing more accessible, limiting the need for a P2P option. Crowdfunding and other products will continue to develop and may obviate the need for a robust P2P lending market. The trajectory of China's P2P lending may be a harbinger of what is to come for all P2P lending. This is natural, however, and should be expected. As FINSIs evolve and grow, they will change, and other innovations may meet needs through more efficient means. Because FINSIs are innovations, they will not remain static.

Note

1 Based on Saucedo and Siegel (2016).

4 Crowdfunding

Crowdfunding is a way of financing projects, businesses, and loans through small contributions from a large number of sources rather than large amounts from a few. The contributions are made directly or through a light-touch platform rather than through banks, charities, or stock exchanges (Agrawal et al. 2014, Baek et al. 2012, Belleflamme et al. 2014, Belleflamme & Lambert 2014, de la Viña & Black 2018, Lombardi et al. 2016, Mollick 2014, Paulet & Relano 2017, Tuomi & Harrison 2017, Turan 2015, Vulkan et al. 2016, Yablonsky 2016). Like P2P lending, crowdfunding is driven by technological advances, and helps to *pool funds* and *move funds*. Like microfinance, crowdfunding helps individuals gain access to resources and investments who were previously denied these services, or traditional financing was inaccessible to them.

Similar to the discussion on P2P lending, crowdfunding has been around for hundreds of years. In Ireland in the 1700s, Jonathan Swift created an "Irish Loan Fund" to support members of his community. The pedestal for the Statue of Liberty in the United States was funded by over 125,000 donors, who contributed either one or five dollars to the cause and were offered small replicas of the statue itself as a reward (BBC 2013). Political campaigns have been crowdfunded for generations, with small donors helping to propel a politician to office. Like all FINSIs, technological advancement supported the acceleration of crowdfunding around the turn of the century. Kiva, previously discussed, helped to crowdfund microfinance loans all around the country. Kickstarter, a crowdfunding platform for artists started in 2009 and popularised the concept of crowdfunding for many, although the term was coined in the United States in 2006. The COVID-19

DOI: 10.4324/9780429297151-5

pandemic further shifted the use of crowdfunding, from a focus on specific projects or businesses to helping individuals make ends meet for basic needs, such as medical bills or accidents (Wang et al. 2022).

There are four different models of crowdfunding: donations, reward, equity, and lending. Lending crowdfunding is essentially P2P lending, discussed in the previous chapter. The other three are as follows.

Donation Crowdfunding

Donation crowdfunding comprises charitable offers to create some valuable social good, without expecting anything tangible in return. Funders have intrinsic and social motivations to participate in this type of crowdfunding and in return they receive intangible benefits. These donations can go to organisations, such as nonprofit organisations and other third-sector entities; individuals, such as those seeking help to recover from an unexpected disaster; or campaigns, such as a politician running for office. Donation crowdfunding helps to expand and accelerate already existing donation efforts, and leverage digital tools to connect to a broader scope of individuals.

Major platforms in this category include GoFundMe, which offers users the chance to fundraise for a wide range of projects, including asking for funds to cover emergency expenses like hospital bills or recovering from a house fire. GoFundMe was used extensively during the initial phase of the pandemic to help those who had lost their jobs to cover their basic needs. Global Giving offers similar fundraising opportunities for organisations to support a campaign for NGOs or other third-sector entities. These typically have a set goal for the campaign, such as funding a new building or training a certain number of people to offer services. Political crowdfunding platforms help facilitate campaign transactions, especially in the United States, which has fewer campaign finance laws. ActBlue is the designated platform for the US Democratic Party, and WinRed has been developed to compete on behalf of Republican candidates.

Reward Crowdfunding

Contributing money to a cause or initiative in return for a valued prize or item, such as a special edition of a product, is reward crowdfunding. These are frequently used for creative or artistic projects. Obtaining the prize is not always the main motivation for contributing and can perhaps be best regarded as a form of sale (tax authorities in the United States and in the United Kingdom levy sales taxes on crowdfunded projects of this kind).

Funders are motivated to participate in this by an intrinsic desire to support the creator or project, and not only receive the reward but also intangible benefits for supporting a project they care about. They are able to access exclusive products by being a part of the campaign (such as a limited edition or signed item), or have unique experiences that they would not otherwise have access to (such as being entered into a lottery to win a chance to meet an artist). For artists and creators, crowdfunding offers a chance to connect more deeply with their fans, as well as access funds with greater creative control.

Major platforms in this category include Kickstarter, which allows users to seek funding for a variety of artistic projects. This platform has been used by a wide spectrum of people, including major Hollywood celebrities seeking direct funding from fans for projects to allow for creative control in exchange for rewards (i.e., the chance to visit a movie set, a personalised note from the artist, etc.). Indiegogo is a similar platform, but focused more on funding ideas or inventions rather than creative projects. Finally, Patreon offers more of a community subscription model, in which artists or other creators receive monthly payments from fans, which receive exclusive products in exchange, such as a monthly newsletter or podcast.

Equity Crowdfunding

In equity crowdfunding small investors can pool funds to invest in a small- or medium-sized enterprises (SMEs), acquiring a stake or equity in the company. This type of crowdfunding may have all of the same rewards and motivations as the other types, with the added benefit of a financial return (Baek et al. 2012, Applegate et al. 2016).

Equity crowdfunding allows for businesses to expand their reach and tap into a new source of capital, sometimes gaining access to funding when they could not before because of a lack of relationship with traditional financial institutions. Equity crowdfunding can also create the intangible benefits of the reward or donation crowdfunding, where investors feel connected to a community and supportive of something they believe in.

Major platforms in this category include StartEngine or SeedInvest, which both offer users the ability to directly invest in a wide variety of startups. Some equity crowdfunding sites offer a more specific service or fill a more narrow niche. Fundii, for example, is a Canadian-based equity crowdfunding company, but seeks to attract capital to Middle Eastern startups and companies. (I sit on the advisory board of Fundii.)

Crowdfunding is a *process* Social Innovation, where investors and entrepreneurs interact through new models of interpersonal relations. Like P2P lending, crowdfunding is driven by *technology* but also helps entrepreneurs and other individuals access capital that they would not otherwise be able to, driven by inaccessibility of financial services and an *incomplete market*. The main innovations of the crowdfunding FINSI are the pooling of *risk capital* in nontraded companies, as well as *web-based platforms*. The innovation in web-based platforms is identical to the discussion within the P2P lending chapter, and the innovation for risk capital primarily involves crowdfunding for business investments. Instead of relying on accredited investors for capital, equity crowdfunding allows businesses to access capital from anyone, anywhere. For example, Crowdcube and Seedrs allow equity investments in nontraded companies, starting at around USD 15. This opens up capital to businesses that would not have other opportunities for investment and scale (Baek et al. 2012, Applegate et al. 2016).

The FINSI Framework

Table 4.1 illustrates the different categories of the FINSI framework as they relate to crowdfunding. They are as follows:

- Main social innovation: Process. Crowdfunding is an innovation in disintermediation, removing the need for traditional financial institutions, similar to P2P lending.
- Level: Incremental. Crowdfunding creates new opportunities to access capital and other forms of fiscal resources, broadening the number of entities who can participate in different types of financing.
- Dimension: Organisation and individual. Crowdfunding operates at the level of the organisation, with donations going to nonprofit

Table 4.1 Crowdfunding

Financial social innovation	Main social innovation	Level	Dimension	Sector	Financial functions	Main drivers
Crowdfunding	Process	Incremental	Organisation and individual	Private (informal)	Moving funds Pooling funds	Incomplete markets Technology

organisations and third-sector organisations, and reward and equity crowdfunding going to private companies.

- Sector: Informal and Private. Like P2P, crowdfunding operates within the private markets as lending platforms, but due to the dis-intermediation can also operate within the informal economy for those without access to traditional financing options.
- Financial functions: Crowdfunding performs the financial functions of moving funds and pooling funds. Funds are moved between individuals and organisations, and pooled when multiple funders support a single project or company.
- Main drivers: Technology has supported the expansion of crowdfunding in recent years, and incomplete markets for financial services have supported further adoption of the FINSI, specifically among businesses looking to grow and scale.

Growth and Scale

Placing an exact number of the crowdfunding market is difficult, as the different types of crowdfunding are expansive and vast. One could argue that the entire P2P lending market sits within the crowdfunding market. Same with political campaigns. Donation crowdfunding is a very different market and mechanism than equity crowdfunding or reward-based crowdfunding. A major donation crowdfunder, GoFundMe, has raised over USD 15 billion since it began in 2010 (GoFundMe 2021). Kickstarter, a major rewards-based plat-form, has over USD 7 billion pledged to projects over its lifetime (Kickstarter, 2023).

Looking specifically at crowdfunding for businesses, the market is around USD 1.6 billion, projected to grow around 16% in coming years. Like all FINSIs, growth is being driven by greater familiarity and comfort with digital technologies, and social media, in particular, is increasingly being used to crowdfund from potential investors. Artificial intelligence (AI) and machine learning are expected to help increase efficiencies within crowdfunding, as well. COVID-19 helped to accelerate the use of crowdfunding for business investments, and governments have started to encourage the development of crowdfunding sites to support scaling of businesses. In 2021, Dubai introduced Dubai Next to support entrepreneurs within the country (Grand View Research 2023).

The Dark Side of Crowdfunding

Crowdfunding suffers from the same moral hazard issues as P2P lending. Those seeking investments and/or donations have an incentive to overrepresent themselves or their businesses. This can lead to suboptimal outcomes or fraud (Wired 2011). CNNMoney found that over 80% of Kickstarter's projects missed their delivery dates (Pepitone 2012). GoFundMe has been used for massive fraud, such as the scheme in the United States in which a couple raised USD 400,000 for a homeless man but kept the money and fabricated their story (Patel 2022).

Like P2P lending, crowdfunding can lead to discriminatory results. An analysis of GoFundMe campaigns found that racial disparities within crowdfunding were exacerbated by the pandemic. Black and Asian individuals had a much more challenging time raising funds than others (Wang et al. 2022).

Implications for Social Outcomes

Crowdfunding, like P2P lending, is focused on process innovations. However, it also has specific applications that can support the sustainable development goals (SDGs). Crowdfunding as a FINSI supports greater financial inclusivity, which can lead to increased business growth and lowered rates of poverty. But, because crowdfunding is an incremental FINSI like P2P lending, these benefits to achieving SDGs will be limited. Businesses can use crowdfunding to reach a level of scale that can receive investment from more traditional investors like venture capitalists (Applegate et al. 2016), and crowdfunding can be a way to increase access to capital for businesses that normally do not receive traditional financing, such as minorities or other marginalised entrepreneurs (Kauffman 2019).

Example: Seedrs

Seedrs is an online equity crowdfunding platform, based in the United Kingdom. It was launched in the 2010s and quickly began acquiring venture capital funding. It helps companies raise investments, and by 2019, it had completed 250 company deals, with about a fifth of these being over £1 million. That same year, they had about 8,000 investor exits on their platform (Reynolds 2020). Changing regulations made it easier for them to help companies raise funds, with the EU increasing the total amount a company could crowdfund to €5 million across all EU member countries. At the end of 2021, Republic, another equity crowdfunding platform based in the United States, acquired Seedrs for USD 100 million. At the time, Seedrs had nearly £2 billion in platform investments, with Republic having around USD 1 billion (Alois 2022).

Sample companies which raised investments on Seedrs include

- This: A company that makes plant-based food, with £17.5 million in annual revenue and 10,000 retail distribution points. The company makes high-fibre, high-protein products that create, on average, half as much CO_2 emissions as chicken, a third as much as pork, and a tenth as much as beef. They appeal to a health- and environment-conscious consumer, capitalising on values-driven consumption trends. This set a target of £2,000,000 but raised double that, for more than £4,000,000 from around 1,800 investors.
- Cuvva: An insurance tech company that offers flexible auto insurance, with over 650,000 customers and more than 1.1 million vehicles insured. Cuvva is an app-based product, leveraging the benefits of fintech to help consumers access its products. It offers policies from 1 hour to 28 days, helping people who are borrowing a car for a short amount of time access additional protection. It is popular with younger drivers, who are more tech savvy and also are more likely to not own their own car and borrow from others. Cuvva set an initial target of £2,500,000, but ended up raising much more: over £4,000,000 from around 3,000 investors.
- FY!: An online marketplace selling homegoods, with sales of over £24 million in 2021. FY! relies on app-based purchases for its customers and also has established itself as a sustainable brand, tapping into conscious consumer trends. The company relies on local production to reduce emissions, and uses minimal packaging with sustainable materials for shipments. FY! has raised several rounds of funding on Seedrs across seven years, totalling over £9,000,000 from around 850 investors (Alois 2022).

Future Trajectories

Crowdfunding, like microfinance, has established itself as an industry, leveraging digital technologies for expansion and growth. As people become more comfortable with digital financial interactions and fintech products, crowdfunding will become more integrated into the financial lives of the average person. Already donation crowdfunding has become more popular since the COVID-19 pandemic, and reward crowdfunding is an established option for connecting with fans and supporting artists. Political campaigns are increasingly relying on digital donations, as are NGOs, offering donors an easier way to support politicians and third-sector organisations they care about.

Equity crowdfunding represents the highest potential for growth within the crowdfunding FINSI. Post-pandemic equity crowdfunding has exploded as an opportunity for funding for startups and other businesses. The amount of equity crowdfunding investments made in early 2023 was twice as high as that made at the end of 2022, with expectations that this growth will continue (Naysmith 2023). This type of financing offers an alternative method for entrepreneurs to fund their companies, establishing somewhat of a solution for those "missing middle" businesses as discussed in Chapter 2, who are too large for microfinance institutions but too small for traditional banks.

While P2P lending may be on the precipice of fading away, equity crowdfunding may be poised for the exact opposite level of growth. Whereas P2P lending suffers from crowding out and moral hazard problems, equity crowdfunding, when done well and regulated appropriately, can offer up an expansion of a type of investing usually only accessible to the very wealthy. In this way, crowdfunding leverages its ability to pool funds and open up new markets for crowdfunding lenders and beneficiaries alike. If the model remains sound, further innovations could make equity investing as commonplace as retail stock investing is now.

5 Mobile Banking

Mobile banking allows customers to make payments and transfer funds through mobile phones without requiring a bank account (Ashta 2017, Rangan & Lee 2012, Jussila 2015). Funds can be quickly and cheaply transferred among mobile users, and then converted into cash at local stores. Doing so engages a broad network of multiple parties, including private citizens, merchants, mobile operators, public authorities, and often banking institutions, although mobile banking is generally designed to replace banks. With the proliferation of smartphones, mobile banking is a common occurrence in day-to-day life, and customers are more and more likely to eschew physical banking locations for mobile or digital transactions.

This chapter will focus most of the discussion of mobile banking specifically as a FINSI in service of the unbanked and the global poor. When mobile phones started to become widespread in developing countries in the early 2000s, it was estimated that around one billion people had access to a mobile phone but did not have a bank account. This represented an amazing opportunity to leverage technological advances to increase financial inclusion for millions of people (Rangan & Lee 2012).

Mobile banking is frequently treated as a subcategory of microfinance. Mobile banking, however, does not offer micro-credit, but *payments*, and specifically leverages the diffusion of mobile phones through technology. Admittedly, the two systems seem to naturally complement each other. For example, *Happy Loans* in India partners with merchant acquirers and money transfer services to identify small enterprises that have reliable cash flows. It then offers them uncollateralised micro-loans to help manage short-term cash

DOI: 10.4324/9780429297151-6

flow mismatches without using previous savings. All this is managed entirely via mobile phone without requiring someone to open a bank account or have a formal relationship with a financial institution. In Bangladesh, as another example, the mobile banking service bKash operates in tandem with BRAC Bank, because the law limits mobile financial services provision to registered financial institutions.

Mobile banking is an *outcome* Social Innovation, resulting in greater financial inclusion for its users. The vast partnerships created by the different actors involved in the payments redefine the entire economic ecosystem of a country, creating *institutional* and *systemic* change. For example, the Kenyan mobile banking service M-PESA (featured later) empowered a vast number of new ventures across what is now dubbed "Silicon Savannah" (The Economist 2012). This financial inclusion helps increase economic growth and raises standards of living.

Financial Institution Mobile Banking vs. Telecom Mobile Banking

Early mobile banking pioneers, like M-PESA and others, proved the value of mobile banking as a way to open up financial services to new markets – namely, the unbanked. Many players, seeing the market potential, decided to compete with early adopters, with traditional financial institutions and even credit card companies like Visa developing their own mobile payment and banking options (Carney 2019). While many individuals do not have access to physical banking branches, especially those who are poor and/or live in rural areas, almost everyone has access to a mobile phone or at least mobile phone coverage, representing a new segment of customers.

Financial institutions entering the mobile banking market, however, have not had as much success as the telecom companies who are the original pioneers of the FINSI. These financial institutions leveraged services beyond just mobile payments, connecting users to existing products rather than designing a new system to meet a market need. Successful traditional financial institutions entering the mobile banking space have partnered with telecom industries or existing mobile payment companies, such as Safaricom's partnership with the Commercial Bank of Africa, or the East African Equity Bank partnering with Airtel's telecom network (Ludwig 2016).

These kinds of innovations and partnerships between telecom companies and traditional banks should be considered separate from recent fintech innovations from traditional financial companies. While the increasing reliance on mobile phones for banking – through apps or other digital interactions – represents a financial innovation, it is not necessarily a FINSI. There is not a social innovation component to the increasing ubiquity of mobile phones being used for financial transactions, as a reliance on digital technology does not necessarily solve a problem for the unbanked. One must still have a bank account to access these products, which just replicates the barrier of having a relationship with a financial institution from the physical space to the digital space.

The FINSI Framework

Table 5.1 illustrates the different categories of the FINSI framework as they relate to mobile banking. They are as follows:

- Main social innovation: Outcome. Mobile banking produces greater financial inclusion for its users.
- Level: Institutional. Mobile banking operates at the level of institutions, creating a new infrastructure for payments and offering financial services for those that are unbanked.
- Dimension: System. Mobile banking creates systemic change, bringing in millions of people into an economic structure that they previously were denied.
- Sector: Private and Third sector. Mobile banking has created a new private market and engages directly with third-sector (NGO) institutions like microfinance to support the transfer of funds.
- Financial functions: Mobile banking supports the transfer of payments as its main innovation.
- Main drivers: Mobile banking is driven by technology and solving for an incomplete market. Mobile payments are made possible by technological advancement and the exclusion of the poor from banking systems has increased its adoption by users.

Mobile phones quickly spread throughout the developing world after they were first introduced in the 2000s. Mobile phone penetration was just at 12% in 2000 but reached 61% in 2008. The fastest growth was

Table 5.1 Mobile (Branchless) Banking

Financial social innovation	Main social innovation	Level	Dimension	Sector	Financial functions	Main drivers
Mobile banking	Outcome	Institutional	System	Private (third sector)	Payment	Incomplete markets Technology

in developing countries, with close to 0% penetration in 2000 rising to almost 50% in 2008. Africa had the highest adoption rate of all the developing regions at the time, with 32%, in contrast to an adoption rate of financial services of 23%. This led to massive growth in mobile banking based on latent demand (Rangan & Lee 2012).

Since the early adoption of mobile phones, mobile banking has exploded, particularly with the advances in smart phone technology. Mobile phones are now everywhere and considered an essential consumer product. Many do not even see their phone as a product but an extension of themselves. Mobile banking as an industry and mobile payments have expanded similarly – the mobile banking industry was valued at around USD 715 million in 2018 and is projected to reach over USD 1 billion by 2026 (Snehal & Onkar 2020). Mobile payments specifically had a market size of about USD 44 billion in 2021 which grew to over USD 55 billion in 2022. Mobile payments are expected to grow to almost USD 600 billion by 2030, driven by rapid adoption of apps like Venmo, Zelle, and others and a transition to an increasingly cashless society (Custom Markets Insights 2022).

The Dark Side of Mobile Banking

Like with all financial transactions, there is opportunity for fraud with mobile banking. Scams can occur through mobile banking and individuals can deceive payees about who they are or their intentions. About a quarter of customers have experienced fraud via a peer-to-peer transfer service, and about 10% of people say they have sent payments to the wrong person via a digital transfer (Carrns 2023). Additionally, an increasing reliance on cashless transactions tends to discriminate against the poor and the elderly, who often do not have the same level of technology as wealthier consumers (Liao 2019).

As products are developed, it has become clear that a "one-size-fits-all" approach does not work with mobile banking. Instead, the system must be designed to meet the needs of the consumer and ensure responsible practices (Voorhies 2016). Many of the concerns of microfinance, with users leveraging the FINSI to go into debt, sometimes at an unsustainable rate, are relevant to mobile banking as well (Lee & Tang 2014). For example, the mobile payment operator bKash recently changed its payment structure, as its users were making high levels of deposits and withdrawals, sometimes multiple times in a day, and sometimes to the same people, which could lead to the accrual of massive fees (Tech Observer 2021).

Other challenges with mobile banking relate to common trends in other financial and digital innovations: regulatory issues around data storage and data disclosure; data sharing agreements between partners interacting within a mobile banking system; reliability and security issues as smartphones become commonplace and users rely more and more on public hotspots that may be less secure (Lee & Tang 2014). All of these issues can be a barrier to further adoption of mobile banking and the expansion of the FINSI.

Implications for Social Outcomes

Like microfinance, the goal of mobile banking is lifting its beneficiaries from poverty through financial inclusion. Because it is a systemic innovation, mobile banking has benefits and implications beyond just that direct outcome. By accessing the financial system, previously unbanked individuals are empowered to engage in all kinds of economic activity. They can more easily start a business, purchase necessary goods or services, or get out of debt.

Mobile phones and mobile banking have been shown to have direct impacts on overall economic growth. Looking at developing countries' economies during the time of mobile phone adoption, Waverman et al. (2005) found that a developing country with 10 more mobile phones per 100 people would have had a 0.59% higher GDP growth than an otherwise identical country. Enriquez et al. (2007) found that wireless access increased GDP in India by 2%, 5% in China, and 7.5% in the Philippines.

These results have broad connections to the sustainable development goals (SDGs): most notably, Goal 1 (ending poverty), but can also support Goal 5 (gender equality), as women with mobile phones are empowered to make financial decisions. The systemic innovation nature of mobile banking makes it difficult to form a direct tie to SDGs, but this FINSI can be used as a tool to further many of the goals and improve social outcomes. A major use for mobile banking is the sending of international remittances, and the mobile banking provider M-PESA (profiled later) states that international remittances contribute to achieving 12 of the 17 SDGs (M-PESA 2023). It has been estimated that 2% of Kenya's population has been lifted out of extreme poverty because of the mobile payment service.

Example: M-PESA[1]

M-PESA is an early pioneer of mobile banking, which began in Kenya in 2007. At this time, only 10% of Kenya's population had access to financial services. Around 80% of the population lived in rural areas and a majority of banks and ATMs were in urban areas. M-PESA began as a way for microfinance organisations to collect loan payments, but they quickly realised that it was being used to send payments instead.

M-PESA works by using "authorised agents" – frequently small mobile phone stores or other retailers such as barbers or butchers – where customers can register for the service. The customer exchanges cash for electronic money with these retailers and then can send it to anyone throughout the country. These transfers are verified by SMS and by using a specific PIN number. Once the electronic money is transferred, the recipient can go to another local authorised agent and exchange the transfer for cash or spend it directly at an M-PESA merchant (M-PESA 2023).

M-PESA is a subsidiary of Safaricom, Kenya's largest telecom provider. When it first launched, customers had access to around 19,000 agents throughout the country where they could send and receive payments. About 300 companies accepted direct payments from M-PESA, without a need to transfer the payments into cash. This innovation ended up being a massive time-saver for Kenyans, as many had to wait in long queues to pay bills. Instead, with M-PESA, they could send funds directly and instantaneously.

The initial launch of M-PESA was incredibly successful and the service grew quickly and exponentially. After two years, about 45% of domestic money transfers went through M-PESA. By 2013, after just five years, the Kenyan government estimated that around 40% of its GDP flowed through M-PESA. By 2016, the service was facilitating millions of transactions each month and had around 25 million active users throughout the country. Now it has over 50 million users making over $500 billion in transactions each year. It also has hundreds of thousands of agents worldwide, no longer limited to just Kenya.

M-PESA soon began to offer and facilitate services beyond mobile payments. Safaricom started to use mobile technology to allow people to open up a "real" bank account through their phone, without the need for a physical branch. This grew to include savings accounts, insurance, and supporting microfinance transactions. Through M-PESA, Safaricom was able to increase financial inclusion among a population of people who had been shut out from traditional financial markets.

All of this success created competitors. Cell service providers Airtel and Orange created their own mobile payment services soon after M-PESA launched, but these never reached the market share of the original offering. M-PESA has maintained around two-thirds of the market share of mobile payments in Kenya. Traditional players were inspired by M-PESA to offer mobile banking services in Kenya and around the world: Visa started to offer mobile banking across Africa and a regional bank, Equity Bank, created its own mobile banking services. What began as a simple innovation in Kenya created a broad, international FINSI market.

Future Trajectories

As mobile payments continue to expand, they are increasingly being subsumed within a broader regime of technologically advanced microfinance offerings. Many mobile banking providers also now offer some level of microcredit, microsavings, or microinsurance (Lee & Tang 2014, Ludwig 2016, Voorhies 2016, Carney 2019). This has expanded access to financial services to a broader population of unbanked individuals, creating greater financial inclusion and the economic benefits that come from more consumers and entrepreneurs entering the formal economy. Like P2P lending, mobile banking may be reaching another phase of its evolution, in which the innovations of this FINSI are fully ingrained in our day-to-day lives. The mobile banking market has become fully established as an alternative financial infrastructure and will likely persist as long as people continue to be reliant on mobile phones.

Note

1 Adapted from Rangan and Lee (2012) and Carney (2019).

6 Impact Investing

Impact investments are "investments intended to create positive social or environmental impact beyond financial return" (O'Donohoe et al. 2010). Impact investing is related to or frequently conflated with terms such as responsible investing, sustainable investing, corporate social responsibility, venture philanthropy, or environmental, social, and governance (ESG) investment criteria (Lanteri 2016, Ferraro & Pathak 2021). Many of these terms connect to a broader movement to use private finance for social outcomes, but impact investments typically deal with the direct investment of funds into social enterprises or businesses that tackle social issues, seeking to help expand those firms and thus their social impact (Lanteri & Perrini 2021).

Like most FINSIs, impact investing has roots deep into history. Many religions prevented the investment of money for a return, which limited the creation of a broad investment market. *The Bible* explicitly prohibits interest, as does *the Qur'an*, which refers to it as *riba* or "increase". As rules and restrictions began to change, religion continued to support and encourage the responsible use of money. Quakers in the 1700s supported slavery abolition by forbidding their members to invest in industries connected to the slave trade. Methodists discouraged congregants from investing in things that could "hurt our neighbour", such as gambling or alcohol. This admonishment extended to things not as tied to *the Bible*, such as the chemical industry with harmful health effects. Islam did not allow any business activity that was considered *haraam* or "forbidden" by *the Qur'an*, which led to the development of an entire Islamic financial services industry. These rules across religions were the foundation of

DOI: 10.4324/9780429297151-7

a broad framework of ethics that are still used by churches and other faith-based institutions today (Ferraro & Pathak 2021).

Modern socially responsible investing began in the 1970s as a way to ensure investments were not doing harm, especially in the context of the environment. One of the major campaigns of this time was the divestment from Apartheid South Africa, where investors of all types, including faith institutions, led a coordinated campaign to choke off funds from the Apartheid government (ICCR, MacAskill 2015). Early firms like Calvert, Domini Investment, Parnassus, and others offered clients services to ensure that their investments were not producing suboptimal social and environmental outcomes, referred to as "negative screens" in the context of socially responsible investing strategies. These same firms soon began experimenting with "positive screens", meaning prioritising firms that were creating social and environmental value rather than just not doing harm (Ferraro & Pathak 2021). This would be the first step towards the field of impact investing as we now know it.

Impact investing in its current form was launched by the Rockefeller Foundation in 2007. Rockefeller enlisted numerous other foundations, NGOs, high-net-worth individuals, development finance institutions, governments, and international organisations attracted by this new way of supporting the causes they care about. In a seminal report by JP Morgan, funded by the Rockefeller Foundation and the Global Impact Investing Network, the authors outlined four principles of impact investing:

- Provide capital: When impact investing first began, investments were either private debt or equity investments in social enterprise companies. JP Morgan expected more publicly traded investment opportunities as the market matured, which became true as more social enterprises became publicly traded companies, although most impact investments remain within the private markets.
- Business designed with intent: The business (fund manager or company) into which the investment is made should be designed with intent to make a positive impact, which differentiates impact investments from investments that have unintentional positive social or environmental consequences.
- To generate positive social and/or environmental impact: Positive impact should be part of the business strategy and measured as part of the investment metrics.

- Expect a financial return: There is a minimal expectation with impact investments for the return of the principal, potentially with some interest (either above or below market rate). This criterion thus excludes donations and philanthropic commitments (O'Donohoe et al. 2010).

Impact investments are an *outcome* Social Innovation, focused on creating a measurable, positive social or environmental impact, and some financial return, either at market level or at *concessionary rates* (Dagger & Nicholls 2016, O'Donohoe et al. 2010, Urban & George 2018). This requires taking equity stakes in or lending money to (*moving funds*) social purpose organisations, instead of giving grants and donations. However, impact investors have a wide spectrum of expectations with respect to the trade-off between impact and return. For example, Vox Capital (profiled later) invests in profitable and fast-growing ventures in different industries, which achieve impact by serving Brazil's poorest citizens, and has a target return of 6% p.a. above inflation. Educate Global Fund, instead, only invests in educational programmes, mostly in East Africa, and aims at capital preservation only, being willing to forgo returns in favour of impact.

The Impact Investment Spectrum

One common debate within impact investing is whether it is considered an "asset class" alongside other asset classes like stocks and bonds, or if firms should incorporate the principles of impact investing across all of their assets. The value of viewing impact investment as an asset class is that investors would then seek to create products in service of impact investing's goal, similar to products in mutual markets or bonds (Lanteri 2016). Impact investors can fall on a broad spectrum from "impact first" to "finance first" investments, in which investors with more interest in impact can focus on breakeven investments or reinvestments of profits back into the funds (see Figure 6.1). Investors focused on a financial return may rely on traditional corporate social responsibility efforts (i.e., divestments from negative impact companies) or through traditional corporate philanthropy. Impact investing can encompass both investments in traditional businesses as well as investments in social purpose businesses with few profits but large social impact.

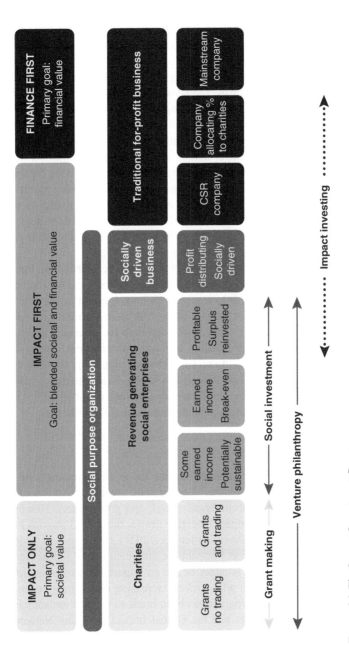

Figure 6.1 The Impact Investment Spectrum.

However, some investors take a full-portfolio approach to impact investing, and leverage all of their investments for impact rather than seeking out a specific subset for impact alone. Investors who approach impact investing more holistically, sometimes called "system-level investors", define an issue or set of issues that they care about, and seek out all avenues to improve the issues, beyond investments in social enterprises. These include direct engagements with companies to set standards or practices, such as "shareholder activism" approaches, in which investors in a public company use their voting power as shareholders to influence decisions within a company towards a social or environmental goal. For example, investors in the mining company Vale leveraged their shareholder power to create new safety standards for workers and the communities around them after a dam run by the company broke and killed over 250 people. This helped set safer standards for the mining industry beyond just the single company (The Investment Integration Project, 2022).

While there is a debate about the differences across the impact investing spectrum, it is a false distinction in some cases. There are many ways for investors to engage with impact investments, whether fully committing their portfolios to 100% investments in social enterprises, leveraging shareholder activism to influence change at companies deemed irresponsible, or relying on traditional responsible investment practices to "do no harm" by not supporting the growth of bad actors. As impact investing grows in popularity and more individuals and companies demand the services, a broad range of options will help support the development and robustness of the industry.

Impact Measurement

An expansion of impact investing has led to an increased need for metrics and frameworks that can illustrate the impact (often called ESG metrics for environmental, social, and governance). However, as the popularity of impact investing grows, these ESG metrics have become less reliable (Perrini & Iantosca 2021). (See The Dark Side of Impact Investing section.) Impact measurement offers a clear framework for determining whether or not an impact investment achieves the outcomes it says it will, such as increasing clean energy or reducing poverty. Much progress has been made to support impact measurement, through tools like the Global Impact Investing Network's database of metrics based on thematic social and environmental areas.

But much more can be done to support the now USD 1 trillion industry of impact investments.

Cole et al. (2022) set out some specific criteria for impact measurement, including:

- Issue or sector, such as affordable housing, off-grid energy, microfinance, or agriculture.
- Target population, such as "base of the pyramid", "emerging middle class", or rural poor. Gender lens investing (in which investments prioritise empowering women) is an example of a strategy defined based on benefits to a specific population.
- Target outcomes, such as improved health or reduced poverty.
- Geography, such as a specific country, region, or even more broadly, emerging markets.

Using these criteria, an impact investment firm could determine a set of metrics that they want to track based on their stated goals, such as the number of women who become employed as a result of the social enterprises within a specific investment portfolio.

The FINSI Framework

Table 6.1 illustrates the different categories of the FINSI framework as they relate to impact investing. They are as follows:

- Main social innovation: Outcome. Impact investing seeks to create social and environmental outcomes by supporting social enterprises or social purpose organisations.
- Level: Institutional. Impact investing is turning into a distinct asset class and even creating a new market for social finance, influencing financial institutions across the globe.
- Dimension: System. Impact investing is creating systemic change, creating a market for investments in social and environmental challenges and helping to scale and grow social enterprises.
- Sector: Private and third sector. Impact investing requires private financial returns; however, many philanthropic (third-sector) organisations have engaged with impact investing to offer concessionary rates to support social impact. Many of these third-sector institutions see public value in supporting the development of impact investing as a private market.

Table 6.1 Impact Investing

Financial social innovation	Main social innovation	Level	Dimension	Sector	Financial functions	Main drivers
Impact investing	Outcome	Institutional	System	Third sector (private)	Moving funds	Incomplete markets

- Financial functions: Impact investing moves funds, directing them to social enterprises, allowing these organisations to expand their reliance to private capital rather than just donations or grants.
- Main drivers: Impact investing is responding to an incomplete market for capital for social enterprises.

Growth and Scale

In 2022, there were USD 1.1 trillion in impact investment assets under management (AUM) worldwide, managed by over 3,000 organisations. This was the first time that impact investments passed the USD 1 trillion mark. This is up from merely USD 8 billion in 2012, an over tenfold increase in a decade (JP Morgan 2013). These assets are controlled by a wide variety of firms: from traditional asset managers to family offices to philanthropic foundations. The majority of the assets are controlled by asset managers (63%), with the next largest category being foundations (11%) (GIIN 2022).

These assets are massively overrepresented in the developed world. Around 40% of the AUM was managed by firms headquartered in the United States or Canada, and just over 50% of the AUM was managed by firms headquartered in Europe. The largest share of AUM outside of these Western countries was in sub-Saharan Africa, with just 2% of total AUM (GIIN 2022).

The Dark Side of Impact Investing

Despite the promise of impact investing, as the asset class has grown in popularity, firms and advisors have jumped on to the trend to try and sell socially and environmentally conscious clients their services. This has led to a dilution of the market and impact investments that are not necessarily robustly tied to the impact they seek to create. This is frequently labelled "greenwashing", in which investments are given a green "coat of paint" so to speak through marketing or messaging that hides sub-optimal social and environmental outcomes. This has led some to claim that impact investing is losing its focus (Casalini & Vecchi 2023) or that impact is in "the eye of the beholder" (Cole et al. 2022). A robust measurement framework can counteract these claims but, unfortunately, even with solid metrics in place, impact measurement, and thus proving out the promise of impact investing, remains challenging. Determining true impact would require a randomised controlled trial study, often considered the "gold standard" of evidence, comparing invested populations to a control group. These studies require a significant commitment of resources and time, which many impact investors do not have (Cole et al. 2022). Impact can also become quite subjective, and different stakeholders may consider different indicators as more significant than others (Harji & Nicholls 2020). For example, the social enterprise may be focused on securing formal sector employment for women to remove them from a more dangerous informal economy, whereas investors primarily care about education attainment for those same women. Integrated reporting, in which a firm or investor would incorporate both sustainability or impact reporting alongside financial reporting, can help to solve these issues and provide a more holistic approach to impact measurement (Granà 2020).

Implications for Social Outcomes

Similar to microfinance, impact investing as a FINSI has a large potential to support the UN's sustainable development goals (SDGs), but as discussed in the previous section, it is hard to make a direct, exact link between impact investments as an industry and the movement towards the SDGs. Individual impact investing firms can make claims towards the SDGs, such as how the microfinance institutions Opportunity International or Maha Agriculture Microfinance make direct connections between their work and progress towards specific SDGs (microfinance could also arguably be considered a subset of impact investing).

However, despite the unsettled nature of impact measurement, SDGs require much more investable capital to reach their 2030 target. Around USD 4.2 trillion in funding is needed to be able to fully resource the SDGs (GIIN 2022). Private capital does not need to fill that entire gap, as governments and NGOs can also support with their own grants and other funding. Yet impact investing, with its current USD 1 trillion capitalisation, can be leveraged to support social enterprises and other markets to provide the funds necessary to meet the SDGs targets.

Example: Vox Capital[1]

Vox Capital was an early pioneer of impact investing, starting in Brazil in 2009. The firm was one of the few funds within the country that was seeking both a financial and social impact. They were responding to a deep need within the country, as at the time, about 80% of its population was living on less than USD 9 a day. Brazil was also one of the most unequal countries in the world. Vox decided to focus its first fund on investing in companies that would help improve the lives of Brazil's poor.

Antonio Ermirio de Moraes Neto is one of the cofounders of Vox, but he was already known throughout the country before he reached working age. His family controls the Votorantim Group, one of the largest industrial conglomerates in Latin America. The family has a history of being humble and giving back, with Antonio's grandfather known for driving an old car, wearing worn-out clothes, and a relentless passion for volunteering. As Antonio explored his career path and his schooling, he decided he wanted to give back and leverage the business tools his family had taught him to do so.

Through his experiences in university, Antonio learned about social businesses and social investing, including impact investing. At the time, very little impact investing was happening in Brazil, and virtually none from local sources. Oikocredit, a subsidiary of a Dutch social investor, made an investment in microfinance in 2003. Antonio decided to bring together some of his cousins in 2007 to use some family funds to invest in social purpose businesses. The family group invested R$ 200,000 in four companies, and ended with an 8% total return.

Antonio took this experience and decided to pursue impact investing not as a side project with family funds but with a completely new entity. He founded Vox Capital with other young, entrepreneurial individuals with connections to Brazil. For their first fund, Vox raised R$ 84 million, much more than they expected. They ended up investing in 20 companies through that first fund, through investments of both debt and equity. Companies were social enterprises that ranged from developing technology to reduce hospital waiting to microcredit firms to online education platforms. Vox wanted to reach 1 million people with its first fund, and they calculated they created 200 jobs and served around 150,000 people daily. The true impact of their investments was difficult to measure, for reasons similar to those discussed earlier, and they applied the lessons learned from their first fund to their second, which they began to raise in 2016. They now have four venture impact funds, including a high-risk fund for the highest impact opportunities. Vox has also developed a nonventure fund to support microenterprises.

Future Trajectories

Key to the expansion of impact investments will be the establishment of a robust impact measurement framework. Without a set of metrics that can show investors the true impact of their investments, both short term and long term, the central promise of impact investing will be questioned and potentially have a detrimental effect on the industry overall (Casalini & Vecchi 2023, Cole et al. 2022). There have been significant efforts undertaken to develop a universal and access-ible impact framework, most notably the Global Impact Investing Network's Impact Reporting and Investment Standards (IRIS), which has hundreds of metrics investors can integrate into their portfolios from 18 impact categories, such as agriculture, education, pollution, and infrastructure. A sub-industry of impact metric managers is growing to support investors, including Novata, which supports pri-vate investment markets; Ecovadis, which supports supply chain management; and Ethos Tracking, which is designed to help impact investors as well as NGOs and other third-sector organisations. Even major corporations like IBM have invested in impact metrics management.

One additional barrier to the expansion of impact investment is the socio-political climate surrounding the emerging FINSI. In the United States, where a large portion of impact investment AUM are housed (GIIN 2022), conservative politicians have used ESG and other impact-first business decisions as a political wedge issue. Many states have restricted a firm's ability to divest from fossil fuels, and companies who make public statements in support of social outcomes are attacked by right-leaning groups. Additionally, social and environ-mental commitments are generally pushed aside during recessionary periods or other economically challenging times (Sullivan et al. 2023), and so if the global economy faces a level of contraction in the coming years, impact investing assets could be transferred to other types of funds, tempering the reach of this FINSI.

Note

1 Based on Battilana et al. (2018).

7 Digital Cryptocurrencies

Digital cryptocurrencies are one of the newest FINSI innovations and rely completely on technology for their existence. Cryptocurrency was first introduced in January 2009 with the development of the Bitcoin protocol and the "mining" for the virtual currency, a type of modern-day gold rush. Originally this FINSI was seen as a very speculative and risky asset, relegated to a specific corner of the internet (Czervionke et al. 2022). By the time I was trying to purchase bitcoin in 2013, cryptocurrency was increasing in popularity. Today, despite some setbacks over the years – notably the breach of bitcoin exchange Mt. Gox in 2014 and the collapse of another major cryptocurrency exchange FTX in 2022 – cryptocurrency has expanded to hundreds of assets with a robust foundational industry.

Digital cryptocurrencies are a *process* Social Innovation, which are distinct from "real world" currencies because they create a decentralised nature *network* of individual participants, who provide the computational power needed to operate the entire system (Ali et al. 2014a, Ali et al. 2014b, Feld et al. 2016, Maull et al. 2017, UK Government Chief Scientific Adviser 2015). The most famous example of digital cryptocurrencies is bitcoin, as it was the first cryptocurrency (Adams et al. 2017, Hileman & Rauchs 2017a). The main motivation for bitcoin was making digital money transactions anonymous and therefore private, like cash transactions. This was not possible before, because the existing *regulations* require that financial intermediaries act as gatekeepers and track all digital money transactions. Early bitcoin users and crypto adopters used the currencies as a way to disrupt the traditional financial system,

DOI: 10.4324/9780429297151-8

but as cryptocurrencies have grown in popularity, many traditional financial institutions have integrated them into their products and services, and consumers have relied on existing relationships and trust within these institutions to engage with the crypto market (Czervionke et al. 2022).

A practical problem with digital money is that it is made of bits, and bits are infinitely replicable. If a currency can be replicated and spent many times over, its value soon becomes zero. To avoid the problem of double spending, digital cryptocurrencies maintain a public ledger of all the transactions, called the *blockchain*, which is the main technological innovation grounded within the cryptocurrency FINSI. Recording transactions on the ledger entails a sophisticated *cryptographic* procedure. This public ledger ensures that the currency can be only spent once, without the oversight of any central authority, therefore radically transforming the relationship between customers and merchants, as well as the operators of the network itself.

The Blockchain

A blockchain is a ledger that records in chronological order all the transactions that have ever been executed within a system. For example, most cryptocurrencies have a ledger that details every transaction ever completed since the creation of the currency. So that one could trace each individual unit of the currency back to the moment of its creation, through each and every transaction, across multiple owners and wallets.

Numerous applications of blockchain beyond payments and across industries are becoming commonplace. The blockchain is now recognised as a general-purpose technology, considered equivalent to a new Internet for its potential to become the technological infrastructure for big data. Because of the proliferation of blockchains, their features can be very diverse (e.g., private or public, permissioned or permissionless, tokenised or not, etc.).

The idea of a blockchain was first described by the elusive inventor of Bitcoin, Satoshi Nakamoto (2008). The name itself comes from the Bitcoin system, where bitcoin transactions are not individually settled, but grouped and processed in blocks of multiple transactions. In order to make the system more resistant to alterations, each block also contains a unique hash number that refers to the previous block.

In this way, each block is linked to the previous and the following one, effectively creating a chain.

Blockchains are typically managed by a peer-to-peer network of computers, called nodes. Each node must adhere to an open-source protocol for the management of the entire blockchain. The protocols that determine how blocks are recorded on the ledger are called consensus, which refers to the notion that the majority of the nodes are in agreement with regards to the correct state of the blockchain. Consensus protocols also create an incentive for nodes to upkeep the network. When a block is recorded on the chain, the protocol creates new cryptocurrency which is given to the node that confirmed the block.

The consensus protocol of Bitcoin is called Proof of Work, which is also known as "mining". So, the nodes are called "mines". Mining consists in finding the solution to a complex mathematical problem. This process consists in a heavy load of "work", which requires very substantial computational power and is therefore costly and energy intensive. The solution can only be found by trial and error, so by generating multiple answers until the right answer is correctly guessed. The right answer is a number lower than the number known as "target hash" (more on this later). When a node finds the solution, all the other nodes in the network verify it. (Interestingly, verifying a correct solution is very easy.) When the majority of the nodes validate the transaction, the block is recorded on the chain and a new block is created. Another increasingly common consensus protocol is called Proof of Stake, first introduced by Peercoin in 2012. Here, the creator of each block is determined randomly, with a system that depends on how much cryptocurrency a node currently owns (or, in some cases, how long they have held it).

Hashing is the process of transforming inputs of various lengths and with various properties into outputs (hashes) with defined length and properties, through a cryptographic procedure that uses a mathematical algorithm. The same input will always result in the same output, but it is impossible to determine the input starting from the hash. Moreover, each hash is unique and cannot be produced using different inputs. These properties help appreciate the difficulty of mining. Finally, even the slightest modification to an input results in a completely different hash.

In exchange for the upkeep of the network, miners are rewarded with newly created bitcoin. New bitcoins are issued at a constant rate, which will accrue to a predetermined maximum amount. This makes the currency intrinsically deflationary. Therefore, its value is expected to grow as the demand for it grows, while supply remains capped. For example, Coinfloor safely stores bitcoin and other cryptocurrencies deposits on behalf of its customers, under the assumption that its value will necessarily grow in the long run.

Consensus protocols and hashing jointly ensure that the blockchain is resistant to any modification of the data, making its records trustworthy without the need of an independent third party. Indeed, once a transaction is recorded in a block, it cannot be modified, without changing its hash, which in turn completely changes all subsequent blocks. Consensus protocols override such modifications because they always enforce the ledger considered trustworthy by the majority.

There is an increasing understanding of the blockchain as a general-purpose technology that can be deployed to increase efficiency and reduce *transaction costs*, including in financial services and particularly in emerging economies. For example, LendLedger is developing a blockchain-based system to facilitate P2P microloans, while disintermediating MFIs.

Crypto as Currency

Cryptocurrencies are a specific subset of virtual currencies, which allow for the exchange of goods or services in the digital sphere. Virtual currencies are a specific subset of all currencies, which can be both physical and digital, regulated and unregulated (see Table 7.1).

Unregulated currencies can be both physical and digital. Unregulated physical currencies are unsanctioned currency, such as

Table 7.1 Types of Currencies

Legal status	Unregulated	Local currencies	Virtual currency
	Regulated	Banknotes and coins	E-money Commercial bank money (deposits)
		Physical	Digital
		Money format	

Source: Adapted from European Central Bank (2012).

a currency developed for local exchanges outside of a national central bank. Unregulated digital currencies are virtual currencies such as cryptocurrencies but also encompass currencies used in video games or other fully digital spaces. Regulated physical currencies are the banknotes and coins that we all use as legal tender in commercial exchanges. Regulated digital currencies are the funds we exchange in a digital space, such as accessing money through our bank's website or app.

There are three types of virtual currencies: Closed currencies, which have no value in the real world and are used to purchase only virtual goods or services, typically within a video game or similar. Unidirectional flow currencies are primarily used to purchase virtual goods or services but can also have some value in the real world, such as Facebook credit or airline miles. Finally, bidirectional flow currencies are established currencies in their own right, with a set exchange rate with other currencies. Digital cryptocurrencies fall into this last category, having full interoperability with the real world (European Central Bank 2012).

The FINSI Framework

Table 7.2 illustrates the different categories of the FINSI framework as they relate to digital cryptocurrencies. They are as follows:

- Main social innovation: Process. Cryptocurrencies have created a new process to exchange value in a digital space, interconnected with the real world.
- Level: Disruptive. The invention of the blockchain and resulting crypto networks has completely upended financial institutions in the last decade, creating a new market and asset class for investment and exchange.
- Dimension: Network. Core to digital cryptocurrencies is the establishment of social networks to maintain a public ledger for each currency. These networks have given the crypto assets their decentralised nature.
- Sector: Informal. Despite the massive growth in recent years, cryptocurrencies remain highly unregulated. This keeps them within the informal sector, even though more regulations are being established to further formalise the industry.

Table 7.2 Digital Cryptocurrencies

Financial social innovation	Main social innovation	Level	Dimension	Sector	Financial functions	Main drivers
Digital cryptocurrencies	Process	Disruptive	Network	Informal	Payment	Technology regulations

- Financial functions: Cryptocurrencies are used for payments, although there are potentially additional applications of the blockchain (see Implications for Social Outcomes section).
- Main drivers: Cryptocurrency began as a way to disrupt the highly regulated nature of existing financial markets, creating more anonymous and decentralised structures.

Growth and Scale

Digital cryptocurrencies have rapidly exploded in growth since bitcoin launched in 2009. The cryptocurrency market is valued at around USD 1 trillion, with 300 million verified cryptocurrency owners. Despite some financial institutions engaging with the markets, the vast majority of these 300 million owners are retail owners, that is, not professional or qualified investors. In 2016, there were only five million owners of cryptocurrencies in the entire world. This number has grown 60-fold since (Czervionke et al. 2022).

With this growth comes an increase in value for the cryptocurrencies. A single Bitcoin was valued at just USD 0.09 when the currency first began. At the end of 2021, it reached its peak, where one Bitcoin was worth almost USD 70,000, a drastic return on investment over just 10 years. The "crypto winter of 2022", where many crypto assets lost value and exchanges like FTX closed, brought prices back down again. Bitcoin ended 2022 priced at around USD 18,000, still a large growth trajectory from its beginnings. Most other cryptocurrencies followed this trend (Edwards 2023).

The Dark Side of Digital Cryptocurrencies

There are two main concerns with cryptocurrencies within the context of FINSIs: That they require massive amounts of energy to produce, causing sustainability concerns, and that the unregulated nature can cause challenges with its original promise of decentralisation.

The process of mining cryptocurrencies and maintaining a public ledger requires massive amounts of energy. Essentially, a cryptocurrency network requires many different computers operating with very high computational levels, which uses a large amount of electric energy to maintain their operations. This causes a lot of power to keep computer servers running. The bitcoin network consumes as much energy annually as the entire country of the Netherlands – and this is just one cryptocurrency of thousands. As sustainability concerns grow within the investing industry (see Impact Investing chapter), this environmental impact becomes a large risk for cryptocurrency overall (Lanteri & Rattalino 2021).

The highly unregulated nature of cryptocurrency also creates a great number of risks. One of the central promises of cryptocurrency is increased decentralisation compared with traditional financial markets, distributing ownership and power across users. However, many of the existing cryptocurrency exchanges and other platforms have operated with a similar level of centralisation as traditional financial institutions, which led to the ignoring of risk exposures and ultimately failure, as was seen in 2022 and other major crypto declines (Ecosystem 2023). Investments in regulatory frameworks and traditional risk management practices can help to solidify the crypto markets and fulfil its promise of centralisation.

Implications for Social Outcomes

Unlike most other FINSIs, cryptocurrencies are still recent innovations. Even FINSIs like impact investing or P2P lending, which accelerated in the last 10–20 years, have roots going back hundreds if not thousands of years. Cryptocurrency, on the other hand, began anew just over 10 years ago. Because of this, much of the long-term impact of cryptocurrencies remains to be seen.

Like other process-oriented FINSIs, cryptocurrency can be leveraged for a social outcome and to contribute to the sustainable development goals (SDGs) based on the existing evidence. Reliance on the blockchain and the public ledger can help to increase trust within decision-making and establish consensus. This allows for greater transparency, more decentralisation, and a diminished reliance on third-party validators of decisions. LendLedger, for example, is using the blockchain to leverage digital data and help borrowers with no financial history access loans. By recording all transactions on the public ledger, the company is able to increase transparency and trust between borrowers and lenders.

Blockchain can also support policymakers in creating more transparent regulations, through the creation of "decentralised autonomous organisations" (DAOs), which are organisations that leverage blockchain technologies to distribute power through a voting process for decision-making managed by a decentralised computer (Zwitter & Hazenberg 2020). Blockchain has been proposed to help set agriculture prices and support contracts, typically a costly process, helping farmers and improving the quality of life (Xiong et al. 2020). Blockchain has also been used to improve credit scores for microfinance institutions and deliver payments to refugees (Wang & De Filippi 2020).

Example: Ripple[1]

Ripple began as a cryptocurrency exchange in 2011, when a group of cryptocurrency pioneers came together to create a new cryptocurrency similar to bitcoin, but one that did not require the same amount of processing power. With this goal in mind, the team developed the XRP ledger and the XRP token. They founded a company to support this ledger, OpenCoin, which quickly rebranded as Ripple as they began to secure venture funding. XRP was the sixth token to be developed, making it one of the first tokens to be created after Bitcoin.

As the company grew and developed, it became clear that their main assets were their holdings of almost 60 billion XRP crypto tokens. These tokens appreciated in value very quickly, mirroring the overall market growth: They were valued at just US 0.05 when first created, but grew to USD 0.36 in May 2017. In 2018, this value skyrocketed to USD 3.60, which made it the highest performing cryptocurrency with a 36,000% gain. However, this valuation was short lived, with a crash in Bitcoin in 2018 pulling down all other cryptocurrencies. Prices fell by 80%.

Despite this decline, the total value of XRP was around USD 20 billion in 2019. However, Ripple's founders now understood the volatility of the market and knew that XRP's valuation could change rapidly. Instead of focusing on building a market for XRP, Ripple decided to do something different and find value for its digital currency beyond just the assessed value. The founders felt that crypto were inherently speculative assets and wanted to create a company that offered more than just a crypto exchange.

They decided to use cryptocurrency as a way to facilitate almost instantaneous cross-border payments. The current international payment system relies on a system driven by large banks and SWIFT (the Society for Worldwide Financial Telecommunications), which is cumbersome and outdated. Transfers can take three to five days and have an error rate of 6%, and cost around USD 40 per transaction. Ripple decided to use cryptocurrency as a way to seamlessly and securely transfer funds anywhere in the world in seconds, disrupting an obsolete system and saving money for clients and consumers.

Future Trajectories

As the most recent FINSI development, there is great potential for cryptocurrency and the blockchain. The market contraction in 2022 and the high-profile failures of some players in the crypto industry were painful and caused many investors to lose a large amount of assets, including retail investors with limited funds. But this contraction may have been a necessary correction of the market, which will lead to greater and more appropriate risk management practices (Ecosystem 2023). Increased regulation may be coming for crypto markets, either through self-regulation or state-sanctioned regulation, which can offer a level of stability for the industry while also supporting the FINSI's decentralised strengths. Some countries are also considering adopting bitcoin or other cryptocurrencies as legal tender. These actions and the development of a valid regulatory framework will counteract negative stories of crypto thefts and other fraud within the industry, which can undermine public trust in the FINSI overall (Bylund 2023). Creating an infrastructure around crypto similar to other asset classes can help to increase stability of the markets and support the longevity of bitcoin and other major cryptocurrencies.

Beyond currency, the blockchain technology has many potential uses that are likely to be explored in the coming years. Decentralised finance (DeFi) leverages the blockchain to allow for financial transactions without the traditional banking methods or models. These transactions use crypto methods and tools for direct connections between individuals, eliminating the need for intermediaries. The decentralised nature of DeFi can be used for things completely unrelated to cryptocurrencies, with the potential for innovation quite high. For example, DAOs rely on blockchain transactions for governance, not currency exchange. DAOs, if implemented correctly, can support a complete restructuring of traditional organisations, creating more transparency and distributed power among staff and employees.

There is huge risk involved, however, as the promise of decentralisation is not always realised within crypto networks, with centralised actors leveraging their power despite proclamations of transparency (Ecosystem 2023). Successfully navigating from hypothetical scenarios of more empowered users and increased trust (Zwitter & Hazenberg 2020, Xiong et al. 2020) to a reality of widespread, integrated use of the blockchain will require much more innovation, experimentation, and adoption of this FINSI.

Note

1 Based on Yoffie and Gonzales (2020).

8 Social Impact Bonds

Government agencies and organisations have always looked for ways to leverage private dollars to produce public outcomes. Public–private partnerships and government contracts with private companies have been used to pave roads, create necessary infrastructure, and develop affordable housing. This trend has accelerated in the past several decades, as governments operate under a neo-liberal approach to governance, seeking out efficiencies and reducing bureaucracy. Privatisation and increased engagement with private sector actors have become a natural outgrowth of a division of labour between sectors, with the government focusing on stewardship of taxpayer dollars, and the private sector focused on the efficient use of those resources (Nicholls & Tomkinson 2015, Vecchi & Casalini 2019).

Social impact bonds are a more recent innovation within the field of public–private partnerships, emerging in the early 2000s alongside other FINSIs like impact investing and P2P lending. Social impact bonds are grounded more in a longstanding tradition of public–private engagements but are a specific type of innovative structured agreement. Social impact bonds are contracts that reward social purpose organisations or social enterprises for achieving agreed-upon, measurable positive impact, like reducing criminal reoffending or improving educational outcomes (Dimitrijevska-Markoski 2016, Lavee et al. 2018, Liebman & Sellman 2013, Gustafsson-Wright et al. 2015, Nicholls & Tomkinson 2015). Often these outcomes are aimed at preventative measures to reduce the burden on the state to support social programmes. Such positive impact generally allows substantial savings for governments that do not need to intervene to address those issues. They can therefore pass parts of these savings on to the social

DOI: 10.4324/9780429297151-9

purpose organisation that made them possible and retain other savings towards taxpayers, creating increased value for all partners involved.

This structure allows social purpose organisations to take risks and attempt innovative social actions, while governments only pay if the innovations actually work. Social purpose organisations raise money from both traditional donors and private investors, with the promise that in case of success they will be paid back.[1] Social impact bonds are often considered part of the impact investing FINSI, because repayment to investors is contingent upon positive impact. For the sake of this book, they are best discussed separately, because they further encourage the emergence of novel multilateral partnerships between governments, social service providers, donor foundations, and investors to tackle social issues, ensuring that each party benefits from the partnership.

Development impact bonds are a specific subset within social impact bonds, which are a distinct category due to the actor delivering repayment for outcomes. The payer for social impact bonds are domestic governments, which reap the benefits of more efficiently delivered social outcomes for the broader taxpayer base. Development impact bonds involve payers that are not domestic governments but a multilateral aid agency or other philanthropic entity. These development impact bonds are usually leveraged for international aid projects in developing countries, with no benefit going to stakeholders of those organisations but instead accruing to the nation or community in which the development impact bond is located (Government Impact Lab).

Social impact bonds are both a *process* and *outcome* Social Innovation, creating a new way of delivering funding to achieving specific impacts within a population or geographic region. The FINSI creates new *networks* from the partnership, encouraging innovation not just in financial payment structures but also in how to deliver and achieve social outcomes.

Outcomes-Based Payment Contract

The main innovation is an *outcomes-based payment contract*, which works as follows:

- A payee or government entity determines what outcomes they want to achieve for a specific social impact bond, such as an increase in graduation rates or employment rates. These outcomes are distinct from "implementation" inputs, activities, or outputs, which are easier to measure but harder to use to determine true impact on a population.

- The payer forms a contract with a social purpose organisation or other social enterprise, setting the terms of the deal, including the specific metrics to measure the chosen outcomes, the timeframe necessary to achieve the outcome, the terms of payment, etc.
- Investors pool money to form the "social impact bond" to offer payment upfront to the payee social purpose organisation. This allows the social purpose organisation to begin implementing their programme. Often these investors are philanthropic organisations or other impact investors.
- With the contract in place, the contracted payee implements their programme in an attempt to achieve the outcomes. If the social outcomes are delivered, then the payer reimburses the investors based on the terms of the contract, giving a return to the investors, and saving taxpayers' money for improved social outcomes (Nicholls & Tomkinson 2014, Gustafsson-Wright 2020).

These contracts can be structured differently, depending on the specifics of the social impact bond, with financial engineering and experimentation supporting increased innovation. For example, establishing different levels of investment within a social impact bond can help reduce risk and bring in additional investors. Creating a "senior" level of investment, where investors are paid out first, would be the least risky and attract more traditional investors, such as Goldman Sachs, which has invested at this level in some social impact bonds. The next level would be paid out next, which would receive more return but would be paid out later, and thus be more risky. The final level would receive the most return but be paid last, increasing the chances of default. A social impact bond could also create a philanthropic level of investment, in which a foundation or other NGO provides an investment with no expectation of return, creating a buffer for other investors in case some of the principal must be forfeited (Kim 2014).

The FINSI Framework

Table 8.1 illustrates the different categories of the FINSI framework as they relate to social impact bonds. They are as follows:

Table 8.1 Social Impact Bonds

Financial social innovation	Main social innovation	Level	Dimension	Sector	Financial functions	Main drivers
Social impact bond	Process Outcome	Institutional	Network	Public Third sector (private)	Pooling funds Risk	Incomplete markets Risk

- Main social innovation: Process and Outcome. Social impact bonds use an innovative financial process to create social outcomes within a region or population.
- Level: Institutional. Social impact bonds create new forms of partnerships between institutions, including governments, social purpose organisations, social enterprises, and investors.
- Dimension: Network. The structured partnership created by a social impact bond creates new networks between the institutions involved and has also begun to create a broader market for investing in the bonds.
- Sector: Public and Third (Private). Social impact bonds operate at the intersection between the public, governmental sector, and the private sector, including nonprofit organisations and other NGOs.
- Financial functions: Social impact bonds help to mitigate risks of innovative and new social programmes by having investors pool funds together and serve as "upfront" payments to social purpose organisations. The risk to the public is diminished because there is no return given without the delivery of the set social outcomes.
- Main drivers: Social impact bonds have responded to an incomplete market for funding social outcomes, as well as the high level of risk that comes from a new social innovation programme.

Growth and Scale

The US-based think tank Brookings maintains a database on all social impact bonds and development impact bonds, which it updates monthly. As of 2023, there were around 240 total impact bonds in 40 countries. The first social impact bond started in 2010 in the United Kingdom, and the growth has been steady since then. The most were in the United States, the United Kingdom, and Portugal, with each having over 15 of the bonds in operation. Around 30 impact bonds operated within developing countries. The total amount of bonds represented around USD 500 million of capital serving around 20,000 beneficiaries (Gustafsson-Wright & Painter 2023).

As of 2020, there were 49 completed social impact bonds. Thirty-three of these led to repayment, and all but one of those led to a full repayment and partial return; the other one only repaid the principal. Fourteen of the bonds are currently under evaluation to determine the outcomes achieved, or have not yet made the outcomes public, and only two of the bonds failed to repay their funds (Gustafsson-Wright 2020).

The Dark Side of Social Impact Bonds

One of the main promises of social impact bonds as a FINSI is to give flexibility to experiment with new innovative social programmes and reduce the risk exposure from the general public from doing so. However, in practice, social impact bonds have mostly focused on established programmes, which generate short-term savings for the partners involved, but limit the ability for greater and more sustainable long-term impact. The focus on established programmes limits the overall risk of the bond, so may be preferred by investors, and can save funds on evaluation costs, which can be quite high (Vecchi & Casalini 2019). However, the lack of experimentation holds back the potential of social impact bonds as an industry.

The nature of evaluative methods also makes it hard to know the true impact achieved from social impact bonds. This is a similar concern as the issues with impact measurement, discussed in The Dark Side of Impact Investing section in Chapter 6, and the proposed solutions in that section are relevant to social impact bonds as well. Brookings did an in-depth analysis of existing social impact bonds and found several concerns with assessing impact of social impact bonds: Even in cases where the established outcomes were achieved through a social impact bond, it is difficult to know whether the mechanism of the bond led to that outcome, because there are no counterfactuals. Rigorous evaluations have not been done of many of the social impact bonds due to their costly nature. There is also concern that the agreed upon metrics to achieve the outcomes of the bonds were sufficiently high enough – that is, it could be the social impact bonds met their outcomes because the targets set were easy to achieve and would have been achieved with or without the intervention. Finally, Brookings points out that few of the social impact bonds have had follow-up research to determine if the outcomes achieved sustained overtime or dissipated over time (Gustafsson-Wright et al. 2020).

Implications for Social Outcomes

Despite these concerns for social impact bonds, there are many implications for social outcomes. By definition, social outcomes are achieved through successful social impact bonds, and development impact bonds in particular have high potential to contribute to the sustainable development goals (SDGs). Two-thirds of social impact bonds have been successful, and only about 4% have been unsuccessful (the remaining are still in progress). The successful social impact bonds have helped to create jobs, keep people out of prison, increase cancer screenings, and enrol girls in school (Gustafsson-Wright 2020). Because of their unique contracting structure, social impact bonds can help achieve any target from the SDGs, depending on which outcomes are agreed upon by the parties involved. Because the social impact bond industry, as compared with broader public–private partnerships, is relatively nascent, more research will need to be done to understand the potential for social impact bonds to influence SDGs and other social outcomes, but they are another valuable mechanism for closing the existing USD 4.2 trillion funding gap needed to achieve the SDGs (GIIN 2022).

Example: Goldman Sachs and Prison Recidivism[2]

One of the first social impact bonds developed in the United States was a structured agreement between Goldman Sachs and a set of partners focused on reducing prison recidivism in Rikers Island, a notorious jail in New York City. Rikers Island can hold as many as 17,000 inmates at one time, and averages about 15,000 daily. Holding people in the jail is an expensive undertaking for taxpayers: at the time the social impact bond was developed in 2012, it cost over USD 160,000 to house an inmate annually. The inmate population was disproportionately made up of minorities and male youths, which had high recidivism rates of over 50%, meaning that more than half of the inmates would return to prison once they left. This led then-Mayor Bloomberg to launch the "Young Men's Initiative" to identify innovative solutions to the incarceration problem.

Social impact bonds were catching on in the United Kingdom, and the model was proposed as a way to fund a more effective and innovative social service solution to the high recidivism rates. Two service-providers – the Osborne Association and Friends of Island Academy – were selected to provide services to around 3,000 inmates, targeting a reduction in recidivism rates. The service-providers were offering an innovative service model using cognitive behavioural therapy as a way to try and change behaviour and reduce reoffending. Using a USD

9.6 million loan from Goldman Sachs, the providers were paid for their services, with the funds being repaid back to the investor if recidivism rates dropped by more than 10%. MDRC, a research organisation, was selected to manage the social impact bond, including administering the funds. Vera Institute of Justice was chosen to manage the evaluation of the project to see if the outcome targets were met.

This social impact bond had the added innovation of a philanthropic guarantee. Bloomberg Philanthropies, the personal foundation of Mayor Bloomberg (who was a billionaire businessman before he became mayor), offered up grant capital to guarantee a certain portion of Goldman Sachs' investment in case of default, lowering the risk for investment. This way, if the outcome targets were not hit, Goldman Sachs would not lose all their funds. The social impact bond was also structured so that if recidivism rates were lower than the expected target, Goldman Sachs would receive additional paybacks. The lower the recidivism, the greater the returns.

Unfortunately, this social impact bond did not meet its targets and was terminated in 2015. The services offered were not shown to reduce recidivism. However, despite the social impact bond not meeting its outcome targets, some still called the programme a success, as it proved out a model for a new and innovative FINSI. The model worked as intended: when the outcome targets were not met, the funds were not repaid, and taxpayer funds were not used for ineffective programmes. The social impact bond structure created an opportunity for innovation in social services, and despite that innovation being unsuccessful, the funding model worked as intended (Cohen & Zelnick 2015).

Future Trajectories

Despite the high-profile termination of the Goldman Sachs social impact bond, most of those executed to date have led to repayment. Over two-thirds of social impact bonds ended in repayment and just about 5% have failed to repay their funds (the remaining repayments are pending evaluations) (Gustafsson-Wright 2020). The last decade of implementing social impact bonds shows that they can be a viable model for the delivering of social services and experimenting with more effective ways to create social outcomes.

Yet the interest in social impact bonds has waned over time. The development of social impact bonds peaked in 2018 at around 50 new contracts that year, and annual growth has fallen since (Gustafsson-Wright 2020). In the last few years, an average of about 15 new social impact bonds are contracted annually (Gustafsson-Wright & Painter 2023).

These diminishing contract amounts are likely a natural effect of decreased interest in social impact bonds. Social impact bonds received a high amount of attention in the early 2010s, especially around the Rikers Island deal, which involved well-known players like Goldman Sachs and Mayor Bloomberg. Some referred to social impact bonds as a "silver bullet" for solving a variety of social problems, from poverty to crime to mental illness (Cohen & Zelnick 2015). No FINSI is likely to be able to deliver on that promise, and more grounded expectations helped to create a more realistic market for social impact bonds.

One major hurdle to an expansion of social impact bonds is related to the challenges outlined in The Dark Side of Social Impact Bonds section and relevant to a similar challenge with impact investing: impact measurement. Evaluating social impact bonds remains costly and each social impact bond must develop its own evaluation framework. This bespoke approach makes it hard to create efficiencies in measurement across social impact bonds. Greater investments in impact measurement will help support the growth and expansion of social impact bonds as it would for the entire impact investing industry. For now, social impact bonds will likely continue at a similar pace.

Notes

1 Social impact bonds do not guarantee regular repayment, or repayment at all, unless the underlying social action is successful. This makes them different from actual "bonds" and more akin to structured products.
2 Adapted from Herrmann et al. (2014).

Conclusion

When I began my financial adventures in Beirut, I did not know how far the path would take me. Nor did I know how long the path stretched behind me. Financial social innovations have roots going back centuries. Since the creation of money, people have been using social means to support the flow of funds and accumulation of wealth. Communities have relied on their members to exchange money and add value. The FINSIs that I reflected on in the terrace of that cafe so many years ago were using technology to accelerate some of the foundational elements of markets and social exchanges.

The far-reaching roots of FINSIs have embedded themselves into our economy in just a few short years. FINSIs are increasingly becoming a fundamental part of our modern way of life, and I do not expect that to change anytime soon. As digital tools become more and more critical to human life, we will see the FINSIs profiled in this book continue to grow and scale, and new ones arise. The framework presented here can help to ground our understanding of FINSIs as the existing ones evolve and new ones are developed.

The discussion of the seven FINSIs profile here suggest four major takeaways:

FINSIs are a unique global phenomenon: As discussed in the introduction, that FINSIs are developing and accelerating at this time in history is not a coincidence. Instead, they are a distinct phenomenon that warrant a set framework of study and exploration. What the seven FINSIs profiled in this book show us is that finance, as an industry, was and is ripe for disruption. Many potential consumers across the world were left out from traditional financial methods yet had needs to be met. Through social innovation, approaches and solutions have

DOI: 10.4324/9780429297151-10

been developed, outside of financial markets to meet the needs of those excluded from that market. Those who were not happy with the status quo sought out ways to change it, creating new opportunities for consumers and bringing in new players to the markets, particularly in developing countries. While some FINSIs were driven by distinctly social needs (such as microfinance, P2P lending, or social impact bonds) and others were more financial in nature (such as cryptocurrency or crowdfunding), all were developed outside the scope of traditional industry players, representing a unified disruption to the industry.

Human ingenuity: Each FINSI is a story of human innovation and human creativity. The approaches and solutions presented here represent the output of an individual or group of individuals identifying a problem that needs to be solved, and then putting in the effort to overcome that problem. Often, these solutions lead to greater efficiencies, in which all parties benefit. For example, social impact bonds, while not technically bonds, are an innovative contracting structure in which different players are able to maximise social and financial returns while reducing risk. Within a social impact bond, the government serves as a good steward of public funds while facilitating necessary funds for social enterprises through an appropriate payment structure for foundation and for-profit investors. Each party to the contract is able to participate and benefit in a way they could not without the social impact bond. Cryptocurrencies, as another example, developed as an alternative system for currency exchange, increasing decentralisation and transparency, and eliminating the need for intermediaries. This has created tremendous value, essentially developing an asset market where there was not one before and allowing for financial transactions without a traditional banking institution.

Rapid growth: FINSIs, while only existing for a relatively short amount of time in comparison with the expansiveness of human commerce, have grown and developed quickly. All saw exponential growth since their first introduction, yet their trajectories have not been homogenous. Some, like cryptocurrencies, emerged at the fringe yet became mainstream. Some were introduced with great fanfare and promise, like social impact bonds, but then levelled out in their adoption. Others experienced high levels of growth, like P2P lending, only to evolve into something else due to their success.

While this book only covered seven FINSIs, it is unlikely that they will be limited to that many for long. Already new concepts are being developed that, in retrospect, could be seen as new FINSIs: Decentralised autonomous organisations, discussed in the chapter on cryptocurrencies, could evolve into a FINSI in its own right. As fintech expands, FINSIs could spring forth where we least expect them. Advances in artificial intelligence could create a new FINSI that we cannot even conceive of today in just a few short years. This book can be revisited, and the framework used as a grounding mechanism to study any new developments within financial and social innovation.

FINSIs as a tool: As noted throughout the book, FINSIs are merely tools without predetermined outcomes. Institutions can leverage FINSIs as they see fit, not necessarily always for positive social outcomes. Microfinance institutions have threatened borrowers, crowdfunders have deceived, and impact investments have been made with dubious intent. Standards and lessons can be collected around each FINSI to posit the best use cases for each one.

This takeaway is especially relevant for the link between FINSIs and the sustainable development goals (SDGs). Because FINSIs are so new, we have not yet figured out the socially acceptable ways to use them. There are still dark sides, as outlined in each chapter. Robust impact measurement can help FINSIs in reaching their full potential. Truly measuring the impact of FINSIs on social outcomes, and the broader SDGs, is not possible in this current moment, although there has been major progress made in recent years. More investment and research in the area of impact measurement will benefit the expansion and evolution of all FINSIs.

Social Innovation remains a "quasi-concept" (Tepsie 2014) requiring more scholarship, and as such, FINSIs also invite more research to dig even deeper. This book laid out a broad framework with components to break down FINSIs and better understand their elements and interactions, which leads to more opportunities to study their components and practical implications. Most of the institutions and individuals operating within FINSIs do not do so using clear lines of division. Instead, many organisations or initiatives encompass many FINSIs at the same time. Microfinance institutions also leverage P2P lending as well as mobile payments, and benefit from impact investing. While teasing out all of these interconnections is beyond the scope of this book, clarity about the seven building blocks is the necessary first step to empower more refined investigations.

While these FINSIs have roots going back generations, they are a distinctly modern phenomenon, reliant on the increasingly powerful digital technologies that are increasingly shaping our lives. Technology is inherently connective, helping to expand FINSIs to a broader group of people and amplify their impact. I hope this book can be a guide as technology continues to advance and brings FINSIs along with it. There is great FINSI potential out there, which can and should be fully leveraged.

References

Acumen Research and Consulting (2023). P2P Lending Market and Region Forecast, 2022–2030.

Adams, R., G. Parry, P. Godsiff, & P. Ward (2017). The future of money and further applications of the blockchain. Strategic Change: Briefings in Entrepreneurial Finance, 26(5): 417–422.

Agrawal, A.K., C. Catalini, & A. Goldfarb (2014). Some simple economics of crowdfunding. National Bureau of Economic Research, w19133.

Akerlof, G.A. (1970). The market for lemons: Quality uncertainty and the market mechanism. Quarterly Journal of Economics, 84: 488–500.

Ali, R., J. Barrdear, R. Clews, & J. Southgate (2014a). Innovations in payment technologies and the emergence of digital currencies. Bank of England Quarterly Bulletin, 54(3): 262–275.

Ali, R., J. Barrdear, R. Clews & J. Southgate (2014b). The economics of digital currencies. Bank of England Quarterly Bulletin, Q3.

Allen, F., A. Demigurc-Kunt, L. Klapper, & M.S. Martinez Peria (2012). The foundations of financial inclusion. Understanding ownership and use of formal accounts. World Bank Policy Research Working Paper, 6290.

Alois, J.D. (2022). Republic – Seedrs ramp up cross-listed securities offerings. Crowdfund Insider. 16 June 2022. www.crowdfundinsider.com/2022/06/192420-republic-seedrs-ramp-up-co-listed-securities-offerings/ (Last accessed 1 June 2023).

Applegate, L., V. Dessain, E. Billaud, & D. Beyersdorfer (2016). Angel investments in Europe and recent developments in crowdfunding. Harvard Business School Technical Note 814-047, March 2014. (Revised May 2016.)

Ariza-Garzón, M.-J., M.-D.-M. Camacho-Miñano, M.-J. Segovia-Vargas, & J. Arroyo (2021). Risk-return modelling in the P2P lending market: Trends, gaps, recommendations and future directions. Electronic Commerce Research and Applications, Volume 49, 101079, ISSN 1567-4223.

Armendariz, B. & J. Morduch (2010). The Economics of Microfinance, 2nd Edition. Cambridge, MA: The MIT Press.

Arti, G., B. Pramod, & K. Vineet (2021). Microfinance Market 2021. Allied Market Research.

Ashta, A. (2017). Evolution of mobile banking regulations: A case study on legislator's behavior. Strategic Change: Briefings in Entrepreneurial Finance, 26(1): 3–20.

Associated Press (2012). Hundreds of suicides in India linked to microfinance organizations, 24 February, 2012.

Austin, J.E., H. Stevenson, & J. Wei-Skillern (2006). Social and commercial entrepreneurship: Same, different, or both? Entrepreneurship Theory and Practice, 30(1): 1–22.

Baek, P., L. Collins, & S. Westlake (2012). Crowding In. National Endowment for Science, Technology and the Arts.

Baek, P., L. Collins, & B. Zhang (2014). Understanding Alternative Finance. National Endowment for Science, Technology and the Arts.

Balogh, S. (2019). LendingClub's turnaround hinges on repackaging consumer loans to win back big investors. Business Insider. https://web.arch ive.org/web/20191206150523/www.businessinsider.com/valerie-kay-chief-capital-officer-lending-club-marketplace-lender-model-2019-12 (Last accessed 1 June 2023).

Basha, S.A., M.M. Elgammal, & B.M. Abuzayed (2021). Online peer-to-peer lending: A review of the literature, electronic commerce research and applications, Volume 48, 2021, 101069, ISSN 1567-4223, https://doi.org/10.1016/j.elerap.2021.101069.

Battilana, J., M. Kimsey, F. Paetzold, & P. Zogbi (2018). Vox Capital: Pioneering impact investing in Brazil. Harvard Business School Case 417-051, January 2017. (Revised November 2018).

BBC (2013). The Statue of Liberty and America's crowdfunding pioneer, 25 April 2013.

BBVA Foundation. Sustainable development goals. www.fundacionmicrof inanzasbbva.org/en/sector-development/sustainable-development-goals/ (Last accessed 30 May 2023).

Bech, M. & R. Garratt (2017). Central bank cryptocurrencies. Bank of International Settlements Quarterly Review, September: 55–70.

Belleflamme, P. & T. Lambert (2014). Crowdfunding: Some empirical findings and microeconomic underpinnings, forum financier. Revue Bancaire et Financiere, 4: 288–296.

Belleflamme, P., T. Lambert, & A. Schwienbacher (2014). Crowdfunding: Tapping the right crowd. Journal of Business Venturing, 29(5): 585–609.

Berdnorz, J. (2023). The History of Peer-to-Peer Lending. P2P Market Data. 2 March 2023.

Boston Consulting Group (2023). Fintech projected to become a $1.5 trillion industry by 2030. 3 May 2023.

Bogen, J. & J. Woodward (1988). Saving the phenomena. Philosophical Review, 97(3):303–352.

Bullough, A., E. Bell, & D. Helm (2015). Opportunity International: Tackling the Rural Hurdle. Thunderbird School of Global Management.

Bylund, A. (2023). The future of cryptocurrency. The Motley Fool. 24 January 2023.

Cajaiba-Santana, G. (2013). Social innovation: Moving the field forward. A conceptual framework. Technological Forecasting and Social Change, 82: 42–51.

Cambridge Centre for Alternative Finance (2016). Cambridge Centre for Alternative Finance. www.jbs.cam.ac.uk/faculty-research/centres/alternat ive-finance (Last accessed 11 April 2016).

Camera, G. (2017). A perspective on electronic alternatives to traditional currencies. Sveriges Riksbank Economic Review, 1: 126–148.

Carney, W. (2019). M-PESA Matters: The Price of Success. Case Development Centre-Rotterdam School of Management.

Carrick-Cagna, A.M. & F. Santos (2009). Social vs. Commercial Enterprise: The Compartamos Debate and the Battle for the Soul of Microfinance. Case Study no. INS105. Fontainebleau, France: INSEAD Case Publishing.

Carrns, A. (2023). Easy to use, mobile payment apps are also easy to misuse. New York Times. 28 January 2023.

Casalini, F. & V. Vecchi (2023). Making impact investing more than just well-meaning capital. Business & Society, 62(5): 911–916. https://doi.org/ 10.1177/00076503221112864.

CGAP (2004). Key principles of microfinance. www.cgap.org/sites/default/ files/CGAP-Consensus-Guidelines-Key-Principles-of-Microfinance-Jan-2004.pdf. (Last accessed 30 May 2023).

Chambon, J., A. David, & J. Devevey (1982). Les Innovations Sociales. Paris: PUF.

Chen, N., A. Ghosh, & N.S. Lambert (2014). Auctions for social lending: A theoretical analysis. Games and Economic Behavior, 86: 367–391.

Christensen, C. (2012). Disruptive innovation explained. https://hbr.org/ video/2226808799001/disruptive-innovation-explained (Last accessed 26 May 2018).

Christensen, C., M. Raynor, & R. McDonald (2015). What is disruptive innovation? Harvard Business Review, December.

Cohen, D. & J. Zelnick (2015). What we learned from the failure of the Rikers Island social impact bond. Non Profit Quarterly. 7 August 2015.

Cole, S., V.S. Gandhi, & C.R. Brumme (2022). Background Note: Managing and Measuring Impact. Harvard Business School Background Note 218-115, April 2018. (Revised June 2022).

Convergences (2019). Microfinance Barometer 2019, 10th Edition. www. convergences.org/wp-content/uploads/2019/09/Microfinance-Barometer-2019_web-1.pdf (Last accessed 30 May 2023).

Crane, D., K. Froot, S. Mason, A. Perold, R. Merton, Z. Bodie, E. Sirri, & P. Tufano (1995). The Global Financial System: A Functional Perspective. Boston: Harvard Business School Press.

Crossan, M.M. & M. Apaydin (2010). A multi-dimensional framework of organizational innovation: A systematic review of the literature. Journal of Management Studies, 47(6): 1154–1191.

Culkin, N., E. Murzacheva, & A. Davis (2016). Critical innovations in the UK peer-to-peer (P2P) and equity alternative finance markets for small firm growth. The International Journal of Entrepreneurship and Innovation, 17(3): 194–202.

Cull, R., A. Demirgüç-Kunt, & J. Morduch (2009). Microfinance meets the market. Journal of Economic Perspectives, 23: 167–192.

Custom Markets Insights (2022). Mobile payments market. https://aws.amazon.com/marketplace/pp/prodview-3kupdv3q76s62#overview (Last accessed 1 June 2023).

Czervionke, C., N. Dienerowitz, W. Lamolo, U. Koyluoglu, K. Lai, L. Sizaret, & M. Wagner (2022). Navigating Crypto: How Financial Intermediaries can Integrate Cryptoassets. Oliver Wyman.

Daggers, J. & A. Nicholls (2016). The Landscape of Social Impact Investment Research. Oxford: Said Business School.

Dawson, P. & L. Daniel (2010). Understanding social innovation: A provisional framework. International Journal of Technology Management, 51(1): 9–21.

de la Viña, L. & S. Black (2018). US equity crowdfunding: A review of current legislation and a conceptual model of the implications for equity funding. The Journal of Entrepreneurship, 27(1): 83–110.

Demigurc-Kunt, A. & L. Klapper (2012). Measuring financial inclusion. The Global Findex Database, World Bank Policy Research Working Paper, 6025.

Deng, J. (2022). The crowding-out effect of formal finance on the P2P lending market: An explanation for the failure of China's P2P lending industry. Finance Research Letters, Volume 45, 102167, ISSN 1544-6123, https://doi.org/10.1016/j.frl.2021.102167.

Dimitrijevska-Markoski, T. (2016). Social impact bonds: A new tool for governance of social programs – Evidence from the UK, USA and Australia. International Journal of Public Policy, 12(3/4/5/6): 261–275.

Dorfleitner, G., E.M. Oswald, & R. Zhang (2021). From credit risk to social impact: On the funding determinants in interest-free peer-to-peer lending. Journal of Business Ethics, 170: 375–400. https://doi.org/10.1007/s10 551-019-04311-8.

Ecosystem (2023). The Degree of Decentralisation in DeFi: Governance, Management, Processing, and Economic Measures across Crypto Networks. Olivery Wyman.

Edwards, J. (2023). Bitcoin's price history. Investopedia. 24 May 2023.

Enriquez, L., S. Schmitgen, & G. Sun (2007). The true value of mobile phones to developing markets. *McKinsey Quarterly*. February 2007.

European Central Bank (2012). Virtual Currency Schemes. Frankfurt, Germany: ECB.

Feld, S., M. Schönfeld, & M. Werner (2016). Traversing bitcoin's P2P network: Insights into the structure of a decentralised currency. International Journal of Computational Science and Engineering, 13(2): 122–131.

Ferraro, F. & R. Pathak (2021). The Responsible Investing Landscape: From SRI through ESG to IMPACT. IESE Publishing.

FINCA (2019). Sustainable Development Goals (SDGs) Contributions to SDG Targets & Indicators. Washington, DC.

Finch, G. & D. Kocieniewski (2022). Big money backs tiny loans that lead to debt, despair and even suicide. *Bloomberg.* 3 May 2022.

Global Impact Investing Network (GIIN) (2022). 2022: Sizing the Impact Investing Market. GIIN.

GoFundMe (2021). Giving report 2021. www.gofundme.com/c/gofundme-giving-report-2021 (Last accessed 1 June 2023).

Government Impact Lab. Impact bonds. https://golab.bsg.ox.ac.uk/the-basics/social-impact-bonds/ (Last accessed 2 June 2023).

Granà, F. (2020). Redesigning Organizational Sustainability through Integrated Reporting. Routledge.

Grand View Research (2023). Crowdfunding Market Size & Share Analysis Report, 2030.

Gustafsson-Wright, E. (2020). What Is the Size and Scope of the Impact Bonds Market?. Washington, DC: Brookings Institution.

Gustafsson-Wright, E., S. Gardiner, & V. Putcha (2015). The Potential and Limitations of Impact Bonds. Washington, DC: Brookings Institution.

Gustafsson-Wright, E. & E. Painter (2023). Social and Development Impact Bonds by the Numbers. Washington, DC: Brookings Institution.

Hamalainen, T. & R. Heiskala (Eds.) (2007). Social Innovations, Institutional Change and Economic Performance: Making Sense of Structural Adjustment Processes in Industrial Sectors, Regions and Societies. SITRA: Edward Elgar.

Harji, K. & A. Nicholls (2020). The imperative for impact measurement. Oxford Answers. www.sbs.ox.ac.uk/oxford-answers/imperative-impact-measurement (Last accessed 2 June 2023).

Harvard Business Review (2012). Disruptive Innovation Explained. www.youtube.com/watch?v=qDrMAzCHFUU (Last accessed 30 May 2023).

Herrmann, J., A. Gurley, J. Ward, & K. Alexander (2014). Goldman Sachs: Determining the Potential of Social Impact Bonds. The Case Center.

Hileman, G. & M. Rauchs (2017a). Global Cryptocurrency Benchmarking Study. Cambridge Centre for Alternative Finance.

Hmayed, A., N. Menhall, & A. Lanteri (2015). Social incubation and the value proposition of social business incubators: The case of Nabad, in: Jamali, D. & A. Lanteri, London, UK: Palgrave Macmillan, pp. 152–172.

Hulme, D., & M. Maitrot (2014). Has microfinance lost its moral compass?. Economic and Political Weekly, 77–85.

ICCR. History of ICCR. www.iccr.org/about-iccr/history-iccr (Last accessed 2 June 2023).

Idrissi, A. (2015). From necessity to opportunity: The case for impact investing in the Arab World, in: Jamali, D. & A. Lanteri, London, UK: Palgrave Macmillan, pp. 178–207.

Jabotinsky, H.Y. & R. Sarel (2022). How crisis affects crypto: Coronavirus as a test case. Hastings Law Journal, 74: 433.

Jamali, D. & A. Lanteri (Eds.) (2015). Social Entrepreneurship in the Middle East, 2 volumes. Palgrave Macmillan.

JP Morgan (2013). Perspectives on Progress: The Impact Investor Survey. GIIN.

Jussila, A. (2015). Mobile Money as an Enabler for Entrepreneurship: Case Eastern Africa. Master's Thesis Aalto University School of Business.

Katsamakas, E. & J.M. Sánchez-Cartas (2022). Network formation and financial inclusion in P2P lending: A computational model. Systems, 10: 155. https://doi.org/10.3390/systems10050155.

Kauffman (2019). Access to capital for entrepreneurs: Removing barriers. www.kauffman.org/wp-content/uploads/2019/12/CapitalReport_042519. pdf (Last access 1 June 2023).

Kewell, B. & P. Ward (2017). Blockchain futures: With or without bitcoin? Strategic Change: Briefings in Entrepreneurial Finance, 26(5): 491–498.

Kickstarter (2023). Stats. www.kickstarter.com/help/stats (Last accessed 1 June 2023).

Kim, J. (2014). Next-generation social impact bonds. *Stanford Social Innovation Review*. 29 December 2014.

Ki-moon, B. (2015). Statement. https://news.un.org/en/story/2015/12/519 172-sustainable-development-goals-kick-start-new-year (Last accessed 26 May 2018).

Kitcher, P. (1989). Explanatory unification and the causal structure of the world, in: Kitcher, P. & W. Salmon (Eds.), Scientific Explanation. Minneapolis: University of Minnesota Press, pp. 410–505.

Knowledge at Wharton (2018). How fintech is transforming microfinance. https://knowledge.wharton.upenn.edu/article/how-is-fintech-transforming-microfinance/. (Last accessed 31 May 2023).

Krlev, G., D. Wruk, G. Pasi, & M. Bernhard (Eds.) (2023). Social Economy Science: Transforming the Economy and Making Society More Resilient. Oxford University Press.

Lanteri, A. (2015). The creation of social enterprises: Some lessons from Lebanon. Journal of Social Entrepreneurship, 6(1): 42–69.

Lanteri, A. (2016). EngagedX: Benchmarking Impact Investments. Hult Publishing, HLT6-27-16-1008C.

Lanteri, A. & F. Perrini (2021). Causal performativity and the definition of social entrepreneurship, in: Ince-Yenilmez, M. & B. Darici (Eds.), Engines of Economic Prosperity. Palgrave Macmillan: Cham.

Lanteri, A. & F. Rattalino (2021). Bitcoin must become more sustainable, for its own good (as well as the planet's). ESCP Impact Paper No. 2021-38-EN.

Lanyon, D. (2021). Exclusive: Zopa exits peer-to-peer lending. Alt Fi, 7 December 2021.

Larsson, O. S., & T. Brandsen (2016). The implicit normative assumptions of social innovation research: embracing the dark side. Social Innovations in the Urban Context, 293–302.

Lavee, D., Y. Kahn, & Y. Fisher (2018). Benefits of social impact bonds in reducing unemployment levels among ultra-Orthodox Jews in Israel. International Journal of Public Policy, 14(3/4): 235–257.

Lee, H. & C. Tang (2014). Experian Microanalytics: Accelerating the Development of Mobile Financial Services in Developing Markets. Stanford Graduate School of Business.

Lerman, R. (2021). Struggling to stay afloat during the pandemic, people turn to strangers online for help. *Washington Post*. 24 April 2021.

Lerner, J. & P. Tufano (2011). The consequences of financial innovation: A counterfactual research agenda. Annual Review of Financial Economics, 3: 41–85.

Liao, A. (2019). Does cashless society discriminate against the poor and elderly? Data Science W231. https://blogs.ischool.berkeley.edu/w231/2019/10/14/does-cashless-society-discriminate-against-the-poor-and-elderly/ (Last accessed 1 June 2023).

Liebman, J. & A. Sellman (2013). Social Impact Bonds. A Guide for State and Local Government. Cambridge, MA: Harvard Kennedy School Social Impact Bond Technical Assistance Lab.

Lombardi, R., R. Trequattrini, & G. Russo (2016). Innovative start-ups and equity crowdfunding. International Journal of Risk Assessment and Management, 19(1/2): 68–83.

Ludwig, S. (2016). Digital Business Transformation in Silicon Valley Savannah. IMD.

MacAskill, W. (2015). Does divestment work? *The New Yorker*. 20 October 2015.

Maha (2021). Impact Report FY 20/21. https://mahamfi.com/wp-content/uploads/2022/11/MAHA-IMPACT-REPORT_21-22.pdf (Last accessed 30 May 2023).

Mäki, U. (2001). Explanatory unification: Double and doubtful. Philosophy of the Social Sciences, 31(4): 488–506.

Maull, R., P. Godsiff, C. Mulligan, A. Brown, & B. Kewell (2017). Distributed ledger technology: Applications and implications. Strategic Change: Briefings in Entrepreneurial Finance, 26(5): 481–489.

McNeill, D. (2006). The diffusion of ideas in development theory and policy. Global Social Policy, 6(3): 334–354.

Merton, R.C. (1992). Financial innovation and economic performance. Journal of Applied Corporate Finance, 4(4): 12–22.

Milana, C. & A. Ashta (2012). Developing microfinance: A survey of the literature. Strategic Change: Briefings in Entrepreneurial Finance, 21(7–8): 299–330.

Mollick, E. (2014). The dynamics of crowdfunding: An exploratory study. Journal of Business Venturing, 29(1): 1–16.

Monetary Authority of Singapore & Deloitte (2017). The future is here. Project Ubin: SGD on distributed ledger. www2.deloitte.com/content/dam/Deloitte/sg/Documents/financial-services/sg-fsi-project-ubin-report.pdf (Last accessed 26 May 2018).

Moulaert, F., D. MacCallum, A. Mehmood, & A. Hamdouch (Eds.) (2014). The International Handbook on Social Innovation. Cheltenham, UK: Edward Elgar.

Moulaert, F., F. Martinelli, E. Swyngedouw, & S. González (2005). Towards alternative model(s) of local innovation. Urban Studies, 42(11): 1969–1990, www.jstor.org/stable/43197218.

M-PESA (2023). M-PESA stories. www.vodafone.com/about-vodafone/what-we-do/consumer-products-and-services/m-pesa#m-pesa-stories (Last accessed 1 June 2023).

Mulgan G., S. Tucker, R. Ali, & B. Sanders (2007). Social Innovation: What it Is, Why it Matters & How it can be Accelerated. Oxford: Skoll Centre for Social Entrepreneurship Working Paper.

Mumford, M.D. (2002). Social innovation: Ten cases from Benjamin Franklin. Creativity Research Journal, 14(2): 253–266.

Murray, R., J. Caulier-Grice, & G. Mulgan (2010). The Open Book of Social Innovation. The Young Foundation.

Najaf, K., R.K. Subramaniam, & O.F. Atayah (2022). Understanding the implications of fintech peer-to-peer (P2P) lending during the COVID-19 pandemic. Journal of Sustainable Finance & Investment, 12(1): 87–102 DOI: 10.1080/20430795.2021.1917225.

Naysmith, C. (2023). Equity crowdfunding sees explosive 119% monthly growth as alternative financing method picks up steam. *Yahoo News.* 5 April 2023.

Nicholls A. & A. Murdock (Eds.) (2012a). Social Innovation: Blurring Boundaries to Reconfigure Markets. Palgrave Macmillan, Hampshire, UK & New York, USA.

Nicholls A. & A. Murdock (2012b). The nature of social innovation, in: Nicholls A. & A. Murdock (Eds.), Social Innovation: Blurring Boundaries to Reconfigure Markets. Palgrave Macmillan, pp. 1–30.

Nicholls A., J. Simon, & M. Gabriel (Eds.) (2015a). New Frontiers in Social Innovation Research. Palgrave Macmillan.

Nicholls A., J. Simon, & M. Gabriel (2015b), Introduction: dimensions of social innovation, in: Nicholls A., J. Simon, & M. Gabriel (Eds.), New Frontiers in Social Innovation Research. Palgrave Macmillan.

Nicholls, A. & E. Tomkinson (2015). The Peterborough Pilot Social Impact Bond, in: A. Nicholls, R. Paton, & J. Emerson (Eds.), Social Finance. Oxford: Oxford University Press.

Nurkse, R. (2009). Trade and development, in: R. Kattel, J. Kregel, & E. Reinert (Eds.). London: Anthem.

O'Connor, T. & H. Wong (2015). Emergent Properties, in: E.N. Malta (Ed.) The Stanford Encyclopedia of Philosophy. https://plato.stanford.edu/entries/properties-emergent (Last accessed 11 April 2016).

O'Donohoe, N., C. Leijonhufvud, Y. Saltuk, A. Bugg-Levine, & M. Brandeburg (2010). Impact Investments. An Emerging Asset Class. New York, NY: J.P. Morgan Global Research.

Opportunity International (2022). 2022 Impact report. https://opportunity.org/news/publications/reports/2022-impact-report (Last accessed 30 May 2023).

Parkinson, C. & C. Howorth (2008). The language of social entrepreneurs. Entrepreneurship & Regional Development: An International Journal, 20(3): 285–309.

Patel, V. (2022). New Jersey man gets 5 years in prison in GoFundMe fraud case. *New York Times*. 7 August 2022.

Paulet, E. & F. Relano (2017). Exploring the determinants of crowdfunding: The influence of the banking system. Strategic Change: Briefings in Entrepreneurial Finance, 26(2): 175–191.

Pepitone, J. (2012). Why 84% of Kickstarter's top projects shipped late. CNN. 18 December 2012. https://money.cnn.com/2012/12/18/technology/innovation/kickstarter-ship-delay/index.html (Last accessed 1 June 2023).

Perrini, F. & A. Iantosca (2021). I rating ESG: Amarli o odiarli? Economia & Management: La rivista della Scuola di Direzione Aziendale dell'Università L. Bocconi.

Phills Jr., J.R., K. Deiglmeier, & D.T. Miller (2008). Rediscovering social innovation. Stanford Social Innovation Review, 6(4): 34–43.

Pol, E. & S. Ville (2009). Social innovation: Buzz word or enduring term. The Journal of Socio-Economics, 38: 878–885.

Postelnicu, L., N. Hermes, & A. Szafarz (2014). Defining social collateral in microfinance group lending, in: Mersland, R. & R.Ø. Strøm (Eds.), Microfinance Institutions. Palgrave Studies in Impact Finance. Palgrave Macmillan: London.

Ramani, S.V. & M. Vivekananda (2014). Can breakthrough innovations serve the poor (bop) and create reputational (CSR) value? Indian Case Studies, Technovation, 34(5–6): 295–305.

Rangan, V.K. & K. Lee (2012). Mobile Banking for the Unbanked, Harvard Business School Case, 511–049.

ResponsAbility (2014). Microfinance market outlook 2015. www.sustain ablefinance.ch/upload/cms/user/2014_11_Microfinance-Market-Outlook-2015-ResponsAbility-EN.PDF (Last accessed 11 April 2016).

Reynolds, J. (2020). Fintechs help deals on Seedrs jump nearly 50 percent. AltFi. www.altfi.com/article/6055_fintechs-help-deals-on-seedrs-jump-nea rly-50-percent (Last accessed 1 June 2023).

Rykiel, E.J. (1996). Testing ecological models: The meaning of validation. Ecological Modelling, 90: 229–244.

Santos, F.M. (2012). A positive theory of social entrepreneurship. Journal of Business Ethics, 111(3): 335–351.

Saucedo, M. & R. Siegel (2016). Lending Club. Stanford Graduate School of Business.

Seedco Policy Center (2007). The Limits of Social Enterprise: A Field Study & Case Analysis. New York: Seedco.

Sen, A. (1993). Capability and well-being, in: Nussbaum, M. and A. Sen (Eds.), The Quality of Life. Oxford: Clarendon Press, pp. 30–53.

Sen, A. (1999). Development as Freedom. New York: Knopf.

Sfeir, S. (2015). Incubators and Funding Institutions in Lebanon: An Infrastructure for Successful Social Entrepreneurs, in: Jamali, D. & A. Lanteri (Eds.), pp. 173–195.

Snehal, M. & S. Onkar (2020). Mobile Banking Market: 2020. Allied Market Research.

Sullivan, E., L. Castrique-Meier, & S. Pickett (2023). By Staying Committed to Social Impact in Times of Uncertainty, Companies Can See Long-Term Benefit. FSG. www.fsg.org/blog/by-staying-committed-to-social-imp act-in-times-of-uncertainty-companies-can-see-long-term-benefit/ (Last accessed 1 June 2023).

Suryono, R. R., B. Purwandari, & I. Budi (2019). Peer to peer (P2P) lending problems and potential solutions: A systematic literature review. Procedia Computer Science, 161: 204–214, ISSN 1877-0509, https://doi.org/10.1016/j.procs.2019.11.116.

Tech Observer (2021). Bangladesh: bKash cuts cash out charge. 4 October 2021. https://techobserver.in/2021/10/04/bangladesh-bkash-cuts-cash-out-charge/ (Last accessed 30 May 2023).

Tepsie (2014). Doing Social Innovation: A Guide for Researchers, Project: The Theoretical, Empirical and Policy Foundations for Building Social Innovation in Europe (TEPSIE), European Commission – 7th Framework Programme. Brussels: European Commission, DG Research.

The Economist (2012). Upwardly mobile. Kenya's technology start-up scene is about to take off. *The Economist*. 25 August 2012.

The Investment Integration Project (2022). Systemic stewardship: Investing to address income inequality. https://tiiproject.com/wp-content/uploads/2022/01/TIIP-Stewardship-Final.pdf (Last accessed 1 June 2023).

Tufano, P. (2003). Financial innovation, in: Constantinides, G.M., M. Harris, & R.M. Stulz (Eds.), Handbook of the Economics of Finance, volume 1 (part A): pp. 307–335, New York, NY: Elsevier.

Tuomi, K. & R. Harrison (2017). A comparison of equity crowdfunding in four countries: Implications for business angels. Strategic Change: Briefings in Entrepreneurial Finance, 26(6): 609–615.

Turan, S.S. (2015). Financial innovation – Crowdfunding: Friend or foe? Procedia – Social and Behavioral Sciences, 195: 353–362.

Turner Lee, N. (2018). Detecting racial bias in algorithms and machine learning. Journal of Information, Communication and Ethics in Society, 16(3): 252–260. https://doi.org/10.1108/JICES-06-2018-0056

UK Government Chief Scientific Adviser (2015). Distributed Ledger Technology: Beyond Blockchain. London: Government Office for Science.

United Nations (2015). Resolution adopted by the General Assembly on 25 September 2015. www.un.org/ga/search/view_doc.asp?symbol=A/RES/70/1&Lang=E (Last accessed 26 May 2018).

Urban, B. & J. George (2018). An empirical study on measures relating to impact investing in South Africa. International Journal of Sustainable Economy, 10(1): 61–77.

Vecchi, V. & F. Casalini (2019). Is a social empowerment of PPP for infrastructure delivery possible? Lessons from social impact bonds. Annals of Public and Cooperative Economics, 90: 353–369. https://doi.org/10.1111/apce.12230.

Voorhies, R. (2016). Mobile Phones Promise to Bring Banking to the World's Poorest. Harvard Business Review.

Vulkan, N., T. Åstebro, & M.F. Sierra (2016). Equity crowdfunding: A new phenomena. Journal of Business Venturing Insights, 5: 37–49.

Wang, F., & P. De Filippi (2020). Self-sovereign identity in a globalized world: Credentials-based identity systems as a driver for economic inclusion. Frontiers in Blockchain, *2*, 28.

Wang, J., J. Luo, & X. Zhang (2022). How COVID-19 has changed crowdfunding: Evidence from GoFundMe. Frontiers in Computer Science, 20 May 2022. Sec. Human–Media Interaction, 4. https://doi.org/10.3389/fcomp.2022.893338

Waverman, L., M. Meschi, & M. Fuss (2005). The impact of telecoms on economic growth in developing countries. The Vodafone Policy Paper Series. 3.

Weerakoon, C., A. McMurray, N. Rametse, & H. Douglas (2016). The Nexus between Social Entrepreneurship and Social Innovation, ACERE Conference, Gold Coast. 3–5 February.

Wei, Z. & M. Lin (2016). Market mechanisms in online peer-to-peer lending. Management Science, 63(12): 4236–4257.

Westley, F. (2008). The social innovation dynamic. Social Innovation Generation, University of Waterloo. http://sig.uwaterloo.ca/research-publi cations (Last accessed 12 May 2015).

Wilson, K. (2015). Social Impact Investment: Building the Evidence Base, Preliminary Version. Paris: OECD.

Wired (2011). Is crowdfunding the next big thing or an invitation to digital fraud? www.wired.com/2011/12/crowdfunding-big-thing-fraud/ (Last accessed 1 June 2023).

World Economic Forum (2015). The Future of FinTech A Paradigm Shift in Small Business Finance. Geneva.

Woodward, J. (2014). Scientific explanation, in: E.N. Malta (Ed.), The Stanford Encyclopedia of Philosophy. https://plato.stanford.edu/archives/ win2014/entries/scientific-explanation (Last accessed 11 April 2016).

Wyne, J. (2015). Bridging impact and investment in MENA, in: Jamali, D. & A. Lanteri (Eds.), pp. 154–177.

Xiong, H., T. Dalhaus, P. Wang , & J. Huang (2020). Blockchain technology for agriculture: Applications and rationale. Frontiers in Blockchain, 3:7. DOI: 10.3389/fbloc.2020.00007.

Yablonsky, S. (2016). Crowdfunding innovations. International Journal of Services, Economics and Management, 7(2/3/4): 246–264.

Yoffie, D.B. & G. Gonzalez (2020). Ripple: The business of crypto. Harvard Business School Case 719-506, April 2019 (Revised February 2020).

Young Foundation (2012). Social Innovation Overview, Project: The Theoretical, Empirical and Policy Foundations for Building Social Innovation in Europe (TEPSIE), European Commission – 7th Framework Programme. Brussels: European Commission, DG Research.

Zollo, L., G. Marzi, A. Boccardi, & M. Surchi (2015). How to match technological and social innovation: Insights from the biomedical 3D printing industry. International Journal of Transitions and Innovation Systems, 4(1/2): 80–95.

Zwitter, A. & J. Hazenberg (2020). Decentralized network governance: Blockchain technology and the future of regulation. Frontiers in Blockchain, 3:12. doi: 10.3389/fbloc.2020.00012.

Appendix 1
Published Case Studies to Complement This Book

The research in this book is based on FINSI models in general (Chapter 1) and not specific cases. However, discussing them jointly with illustrative case studies is a good pedagogic strategy. Hence, this appendix recommends cases for each of the FINSIs.

Microfinance is the FINSI with the longest history. As a consequence, there are many good-quality case studies, covering a range of rather specific topics within microfinance. At the time of writing, a search for "microfinance" gave several hundred results at the Case Center (www.thecasecentre.org), Harvard Business Publishing (https://hbsp.harvard.edu), and Ivey Publishing (www.iveycases.com). The range of case studies on other FINSIs is more limited but growing very fast. It would be impossible to review all or even most of them. Instead, I am listing a selected few case studies I have successfully used in combination with the materials of this book for advanced undergraduate classes, with students who had solid foundations of business, including microeconomics and finance, and a confident mastery of English.

Microfinance

- Bullough, A., E. Bell, & D. Helm (2015). Opportunity International: Tackling the Rural Hurdle. Thunderbird School of Global Management, TB0391.
- Du, Y., R.O. Chang, M. Wu, & C. Li (2013). Contrasting China's Yunan Model with Bangladesh's Yunus Model for Microfinance. Ivey Publishing & China Management Case-Sharing Center, 9B13N012.
- Kamath, R. & N. Joseph (2019). Microfinance in India: A Tale of Two Models. India Institute of Management, IMB773.

- Pan, G., B. Lee, & L. Bhattacharya (2022). Daung Capital: Managing the Credit Risk Challenges of Microfinance in Myanmar. Singapore Management University, SMU-22-0006.
- Rangan, K., M. Chu, & T. Gregg (2022, revised 2023). Accion's Fintech Strategy. Harvard Business Publishing, 9-319-091.
- Singh, J. & E. Wu (2018, revised 2019). Kiva's Impact Strategy. INSEAD, 718-0063-1.
- Usher, B. (2020). Water.org: Financial Innovation for Impact. Columbia Business School, CCW200307.

Peer-to-Peer Lending

- Pang, T.Q. & M. Huang (2022). FinVolution. Asia Case Research Center University of Hong Kong, 322-0239-1.
- Piskorski, M.J., I. Fernandez-Mateo, & D. Chen (2009, revised 2011). Zopa: The Power of Peer-to-Peer Lending. Harvard Business Publishing, 9-709-469.
- Saucedo, M. & R. Siegel (2016). Lending Club. Stanford Graduate School of Business, E-597.
- Sieber, S. & A. Rafnsdottir (2016). Marketplace Lending at Funding Circle. IESE Business School, IES-623, SI-196-E.
- Tan, S.L., T. Lim, Y.W. Tok, & T. Chansriniyom (2020). From Crowdfunding to Digital Banking: The Evolution of Funding Societies. Singapore Management University, SMU-20-0014.

Crowdfunding

- Blohm, I., P. Haas, C. Peters, T. Jakob, & J.M. Laimeister (2017). Managing Disruptive Innovation through Service Systems – Crowdlending in the Banking Industry. University of St. Gallen, 317-0281-1.
- Kominers, S.D. & A. Nichifor (2021). Birchal: Equity Crowdfunding in Australia. Harvard Business Publishing, 9-820-116.
- Young, N. & K. Lightstone (2013). Crowdfunding at the Brooklyn Warehouse. Ivey Publishing, W13423.
- Zacharakis, A., G. Quintana, & T. Ripke (2016). Crowdfunding: A Tale of Two Campaigns. Babson College, BAB282.
- Zhang, Y., R.A. Malaga, & E. Nunez (2017). StartupValley: Platform Strategy in Equity Crowdfunding. Ivey Publishing, W17155.

Mobile (Branchless) Banking

- Schiller, M. & P. Brest (2014). Bill & Melinda Gates Foundation and bKash: Investing in the Future of Mobile Payments. Stanford Graduate School of Business, SM229.
- Rangan, V.K. & K. Lee (2010, revised 2012). Mobile Banking for the Unbanked. Harvard Business Publishing, 9-511-049.
- Walske, J., E. Foster, & L. D'Andrea Tyson (2018). Grameen America: The Pivotal Role of Technology in Scaling. UC Berkeley Haas School of Business, B5918.

Impact Investing

- Arjalies, D.L., S. Chen, S. Sathasivam, & A. Newton (2020). Verge Capital: Investing for Social Impact. Ivey Publishing, 9B20M135.
- Battilana, J., M. Kimsey, F. Paetzold, & P. Zogbi (2017, revised 2018). Vox Capital: Pioneering Impact Investing in Brazil. Harvard Business Publishing, 9-417-051.
- Lanteri, A. (2016). EngagedX: Benchmarking Impact Investments. Hult Publishing, HLT6-27-16-1008C.
- Rangan, V.K. & S. Appleby (2013, revised 2017). Bridges Ventures. Harvard Business Publishing, 9-514-001.
- Vecchi, V., F. Casalini, N. Cusumano, & M. Brusoni (2015). Oltre Venture: The First Italian Impact Investment Fund. SDA Bocconi, 315-066-1.
- Wertmen, A. & B. Rostoker (2021). SoLa Impact and the Billion Dollar Social Impact Fund. USC Marshall/Lloyd Greif Center for Entrepreneurial Studies, SCG-592.

Digital Cryptocurrencies

- Alfaro, L., C. Larangeira, & R. Costas (2022). El Salvador: Launching Bitcoin as Legal Tender. Harvard Business Publishing, 9-322-055.
- Allayannis, Y.B. & A. Sesia (2023). Crypto Winter Buries Celsius Network and Batters DeFi. Darden Business Publishing, UVA-F-2027.
- Barinaga, E. & J. Ocampo (2019). FairCoop: The Global Cooperative and its Collaborative Cryptocurrency. Copenhagen Business School, 819-0071-1.

- Di Maggio, M. & W. Sha (2021). Coinbase: The Exchange of the Cryptos. Harvard Business Publishing, 9-222-044.
- Yoffie, D. B. & G. Gonzalez (2020). Ripple: The Business of Crypto. Harvard Business Publishing, 9-719-506.
- Vergne, J.P. & B. Burke (2015). BitGold: Turning Digital Currency into Gold? Ivey Publishing, W15608.

Social Impact Bond

- Hoffman, A., J. Herrmann, A. Gurley, J. Ward, & K. Alexander (2014). Goldman Sachs: Determining the Potential of Social Impact Bonds. WDI Publishing, University of Michigan, W93C75.
- Levy, D. & P. Varley (2014). Betting Private Capital on Fixing Public Ills: Instiglio Brings Social Impact Bonds to Colombia. Harvard Kennedy School, KS1003.
- Quelch, J.A. & M. Rodriguez (2014, revised 2017). Fresno's Social Impact Bond for Asthma. Harvard Business School & Harvard T.H. Chan School of Public Health, 9-515-028.
- Rangan, V.K. & L.A. Chase (2013, revised 2015). Massachusetts Pay-for-Success Contracts: Reducing Juvenile and Young Adult Recidivism. Harvard Business School, 9-514-061.

Appendix 2

Theoretical Validation

By mapping the seven FINSIs onto the proposed framework, it becomes evident that they represent every possible variation of each of the elements in the framework (Table A2.1). So, all the characteristics identified are indeed relevant and useful for classifying FINSIs. This validates the framework.

This study does not attempt to test a theory. So, the claim that the framework is *validated* merely means that it "possesses a satisfactory range of accuracy consistent with the intended application […] within its domain of applicability" and "not that it embodies any absolute truth" (Rykiel 1996, p. 233). The main result of this research is an entirely new classification system for FINSIs.

The second relevant result is that these different FINSIs, which have so far been considered disjoint, can instead be regarded as distinct manifestations of a common underlying phenomenon of Social Innovation in the field of finance. Unificatory explanations, capable of accounting for a large set of phenomena, have often been regarded as desirable achievements in the history of scientific knowledge, under the tacit belief that they must uncover some ultimate principle or fact about the world (Bogen & Woodward 1988, Kitcher 1989, Mäki 2001, Woodward 2014). Often, they also display parsimony, the epistemic virtue of explaining much by little.

Theories perform one or more of three functions: they describe, explain, and predict phenomena. Admittedly, as employed in this research, the framework is largely descriptive and does not venture into strong causal explanations of FINSIs. Nonetheless, "[s]ome kinds of unification consist in the creation of a common classificatory scheme or descriptive vocabulary where no satisfactory scheme previously existed, as when early investigators like Linnaeus constructed comprehensive and principled systems of biological classification" (Woodward 2014).

The systematic mapping completed for this book results in a novel framework to classify FINSIs, with several epistemic virtues and particularly the unification of different FINSIs under one theoretical account. What else is to be gained by this theoretical unification, beside theoretical merit badges?

On the one hand, Social Innovation scholars usually bundle case studies from very different fields (e.g., distance learning, emissions trading, and fair trade) in their research. This is suboptimal because, clearly, these cases differ under several respects. So, it may be difficult to draw strong or robust conclusions from such a research strategy. By narrowing down the field of investigation to finance alone, more pointed comparisons between cases become possible. This allows clearer findings that can improve our understanding of the concept of Social Innovation at large. On the other hand, it invites the study of FINSIs with the emerging rich scholarly toolkit of Social Innovation, which can illuminate several aspects of the process of value creation and of the path to innovation (social, financial, or both), and business models, particularly hybrid ones. These two are additional, albeit important, theoretical contributions.

Table A2.1 Framework Validation

Financial social innovation	Main social innovation	Level	Dimension	Sector	Financial functions	Main drivers
Microfinance	Process Outcome	Disruptive	System	Private (third sector)	Extract information Asymmetric information	Incomplete markets Transaction costs Asymmetric information
P2P lending	Process	Incremental	Individual Organisation	Informal, private	Moving funds Pooling funds	Technology
Crowdfunding	Process	Incremental	Organisation	Private (informal)	Moving funds Pooling funds	Incomplete markets Technology
Mobile banking	Outcome	Institutional	System	Private (third sector)	Payment	Incomplete markets Technology
Impact investing	Outcome	Institutional	System	Third sector (private)	Moving funds	Incomplete markets
Digital cryptocurrencies	Process	Disruptive	Network	Informal	Payment	Technology Regulations
Social impact bond	Process Outcome	Institutional	Network	Public, third sector (private)	Pooling funds Risk	Incomplete markets Risk

Glossary

Agency problems   refer to a conflict of interest between the owner of a business (or principal) and the person that acts on his behalf (or agent).

Algorithm   is a process or a set of instructions to perform a task or solve a problem.

Asymmetric information   occurs when only one party to a transaction possesses some critical information that affects the value or the viability of the transaction.

Blockchain   is a distributed ledger, managed by a network through an open-source system based on public cryptography and a **Consensus** protocol (see below).

Collateral   is an asset offered as a guarantee.

Concessionary rates   are interest rates below market rates for a given risk level.

Consensus   is the **Algorithm** (see above) used by cryptocurrencies to validate transactions. The two main types of consensus are Proof of Work and Proof of Stake.

Credit scoring systems   reflect the riskiness of lending money to a borrower.

Cryptography   is a set of techniques for protecting data so that they can be accessed and modified only by authorised parties. It is used for data storage, transmission, and authentication.

Currency   is generally accepted **Money** (see below). Fiat money is issued by an authority and is not backed by a physical commodity. Legal tender is money that must be accepted as payment by law.

Disintermediation   is the processing of transactions without the involvement of intermediaries (or middlemen).

Dynamic incentives   encourage borrowers to repay debt through the promise of having access to further loans.

Equity   is any security, usually a stock or share, that represents ownership.

Financial inclusion   means access to financial products and services that are affordable. It is often used to refer to giving access to individuals and organisation that do not have access to traditional banking services and are therefore referred to as the "unbanked".

Financial innovation is "the act of creating and then popularising new financial instruments as well as new financial technologies, institutions and markets" (Tufano 2003, p. 310).

Fintech is "the use of technology and innovative business models in financial services" (World Economic Forum 2015). This is the most comprehensive definition. Other definitions limit fintech to technologies or the industry resulting from their application.

Group lending replaces individual **Collateral** (see above) with a collective agreement to repay the debt owed by each member of a group, where peer pressure acts as an incentive to repayment.

Hash is a mathematical function that transforms inputs of any length and with any properties into outputs (hashes) with defined length and properties through a **Cryptographic** (see above) procedure.

Information asymmetry occurs when a party to a transaction has greater or more accurate knowledge than the other parties.

Money is a good that performs three functions: it serves as a store of value, as a means of exchange, and as a unit of accounting. See also **Currency**.

Moral hazard is the risk that a party to a transaction has an incentive to behave in a way that increases the risk of the other parties.

Outcome-Based Payment Contract is a contract specifically designed to establish a payment scheme that supports the creation of social outcomes. They usually involve a social service provider which receives upfront payment, a set of investors who provide that payment, and a government entity to repay the investors, with a return, once the outcomes are achieved.

Platform is a business model that facilitates exchanges between two or more parties.

Risk capital (see **Equity**).

Security is a contract or instrument that represents financial value. For example, a share or a bond.

Social innovation comprises significant and lasting changes in a society or in its structures, regulations, norms, and values, such that the society's performance in satisfying the economic and social needs of its members is enhanced, at the individual, collective, or system level.

Social purpose organisation is an organisation whose main purpose is to create social value or address a social issue.

Transaction cost is any cost incurred in order to successfully conclude a market transaction. For example, gathering accurate information and negotiating and enforcing contractual terms.

Unbanked (see **Financial inclusion**).

Unsecured loan is a loan offered in the absence of a **Collateral** (see above).

Index

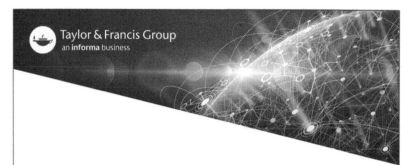